Stated Preference Methods Using R

Chapman & Hall/CRC
The R Series

Aims and Scope

This book series reflects the recent rapid growth in the development and application of R, the programming language and software environment for statistical computing and graphics. R is now widely used in academic research, education, and industry. It is constantly growing, with new versions of the core software released regularly and more than 5,000 packages available. It is difficult for the documentation to keep pace with the expansion of the software, and this vital book series provides a forum for the publication of books covering many aspects of the development and application of R.

The scope of the series is wide, covering three main threads:

- Applications of R to specific disciplines such as biology, epidemiology, genetics, engineering, finance, and the social sciences.
- Using R for the study of topics of statistical methodology, such as linear and mixed modeling, time series, Bayesian methods, and missing data.
- The development of R, including programming, building packages, and graphics.

The books will appeal to programmers and developers of R software, as well as applied statisticians and data analysts in many fields. The books will feature detailed worked examples and R code fully integrated into the text, ensuring their usefulness to researchers, practitioners and students.

Published Titles

Stated Preference Methods Using R, *Hideo Aizaki, Tomoaki Nakatani, and Kazuo Sato*

Using R for Numerical Analysis in Science and Engineering, *Victor A. Bloomfield*

Event History Analysis with R, *Göran Broström*

Computational Actuarial Science with R, *Arthur Charpentier*

Statistical Computing in C++ and R, *Randall L. Eubank and Ana Kupresanin*

Reproducible Research with R and RStudio, *Christopher Gandrud*

Introduction to Scientific Programming and Simulation Using R, Second Edition, *Owen Jones, Robert Maillardet, and Andrew Robinson*

Displaying Time Series, Spatial, and Space-Time Data with R, *Oscar Perpiñán Lamigueiro*

Programming Graphical User Interfaces with R, *Michael F. Lawrence and John Verzani*

Analyzing Baseball Data with R, *Max Marchi and Jim Albert*

Growth Curve Analysis and Visualization Using R, *Daniel Mirman*

R Graphics, Second Edition, *Paul Murrell*

Multiple Factor Analysis by Example Using R, *Jérôme Pagès*

Customer and Business Analytics: Applied Data Mining for Business Decision Making Using R, *Daniel S. Putler and Robert E. Krider*

Implementing Reproducible Research, *Victoria Stodden, Friedrich Leisch, and Roger D. Peng*

Using R for Introductory Statistics, Second Edition, *John Verzani*

Dynamic Documents with R and knitr, *Yihui Xie*

Stated Preference Methods Using R

Hideo Aizaki
Tomoaki Nakatani
Kazuo Sato

CRC Press
Taylor & Francis Group
Boca Raton London New York

CRC Press is an imprint of the
Taylor & Francis Group, an **informa** business
A CHAPMAN & HALL BOOK

First published in paperback 2024

First published 2015 by Chapman & Hall/CRC Press
2385 NW Executive Center Drive, Suite 320, Boca Raton FL 33431
4 Park Square, Milton Park, Abingdon, Oxon, OX14 4RN

CRC Press is an imprint of Taylor & Francis Group, an Informa business

Publisher's Note
The publisher has gone to great lengths to ensure the quality of this reprint but points out that some imperfections in the original copies may be apparent.

Library of Congress Cataloging-in-Publication Data

Aizaki, Hideo.
Stated preference methods using R / Hideo Aizaki, Tomoaki Nakatani, Kazuo Sato.
pages cm. -- (Chapman & Hall/CRC the R series)
"A CRC title."
Includes bibliographical references and index.
ISBN 978-1-4398-9047-9 (hardcover : alk. paper) 1. Decision making--Data processing. 2. Probabilities--Data processing. 3. R (Computer program language) I. Nakatani, Tomoaki, 1969- II. Sato, Kazuo. III. Title.

QA273.A5.A39 2014
519.5'202855133--dc23 2014019441

ISBN: 978-1-4398-9047-9 (hbk)
ISBN: 978-1-03-292447-2 (pbk)
ISBN: 978-0-429-06569-9 (ebk)

DOI: 10.1201/b17292

Visit the Taylor & Francis Web site at
http://www.taylorandfrancis.com

and the CRC Press Web site at
http://www.crcpress.com

Contents

List of Figures

List of Tables

Preface

This book is an introduction to stated preference (SP) methods using R. SP methods are a family of survey methods measuring people's preferences based on decision-making in hypothetical choice situations. The alternatives from which respondents must choose may be commercial goods/services, environmental policy options, travel modes, and others, which show that the methods can be widely used in a variety of disciplines. Although there are many variations of SP methods, we focus on the core ones: contingent valuation (CV), discrete choice experiments (DCEs), and best–worst scaling (BWS). Over the last couple of decades, SP methods have attracted considerable attention among theoretical and empirical researchers in a number of research fields. Yet, when looking at the facilities for empirical applications of SP techniques, there seem to be few coherent resources for statistical computing. Therefore, while giving introductory explanations on the SP methods, this book collates the information on existing R functions and packages as well as those prepared by ourselves.

We have chosen R as a platform for statistical analyses because it is free and open-source. The term "free" means that the monetary cost of obtaining R is negligible if one has access to the Internet, while "open source" indicates that the source code is available to the public under the Free Software Foundation's GNU General Public License. The latter meaning implies that we can read the actual code, which can assist in better understanding the theory and calculation processes of statistical methods. In addition to the fact that R has more than 5,000 user-contributed packages, many books and websites on R are available in a variety of languages. Accordingly, R enables us to learn statistical techniques such as SP methods with limited cost. Furthermore, in this book, we use two or more example datasets to give a clearer picture of empirical applications in each method with R. Examples of CV include data from well-known environmental valuation studies such as the *Exxon Valdez* oil spill in Alaska. To explain DCEs, synthetic datasets are used for three examples related to food marketing and environmental valuation, while the two examples illustrating BWS refer to valuing agro-environmental and food issues. All the example datasets and code are also available to the public allowing readers to easily reproduce working examples. Supplementary materials to the book will be made accessible through the book's web page at http://www.agr.hokudai.ac.jp/spmur/.

The primary audience consists of senior undergraduate and graduate students who are planning to conduct applied research of SP methods in economics and market research. This book is also suitable as a textbook for introductory-level hands-on courses or as supplemental reading in introductory-level courses in these research fields. Considering these points, the chapters covering CV, DCEs, and BWS begin by touching on a minimum amount of theory, followed by examples based on actual or hypothetical empirical studies. Although the examples are limited to those in agricultural/environmental economics owing to the expertise of the authors, we hope that these examples can provide beginners attempting SP methods in wider fields with a good foundation in which they can enjoy learning the SP methods.

While preparing the manuscript over the past couple of years, we have benefited in many respects from numerous individuals, whose support was indispensable in writing this book and developing the packages used therein. First of all, we would like to thank the R Core Team for their unlimited contribution to the provision of a stable R environment. The efforts of all the authors/maintainers of the packages mentioned in this book should also be acknowledged. We are grateful to the series editor of this R Series and the anonymous assessors of our project proposal for their constructive feedback. We would like to acknowledge the comments and suggestions from reviewers of earlier versions of the manuscript. Empirical data on CV studies were made available to us by the kind permission of Professors Richard T. Carson, Bengt Kriström, and John Whitehead. These data are included in the **DCchoice** package. We are also obligated to the users of our packages for their feedback. Our special thanks go to the editorial team of Taylor & Francis/CRC Press, especially Rob Calver. This book would not exist without his encouragement and support throughout this project, as well as his discovery of us.

Materials from this book have been presented at useR! 2013, the R user conference, in Albacete, Spain, and the Japanese R user conferences in 2011 and 2013 at the Institute of Statistical Mathematics, Japan. Comments and suggestions from the delegates at these events are acknowledged. The work of Tomoaki Nakatani was partly supported by the JSPS Grant-in-Aid for Scientific Research (B) 25292131.

Hideo Aizaki
Tomoaki Nakatani
Kazuo Sato

Authors

Hideo Aizaki is an associate professor of the research faculty of agriculture, Hokkaido University, Japan. He received his PhD (agricultural economics) from Hokkaido University in 2000. He worked as a senior researcher at the Institute for Rural Engineering, National Agriculture and Food Research Organization, Japan. His research interests include agricultural and rural development, and consumer preferences for food safety. He is the author of R packages **support.CEs**, **support.BWS**, and **mded**.

Tomoaki Nakatani is an associate professor of agro-food economics and statistics in the research faculty of agriculture, Hokkaido University, Japan. He received a PhD (economic statistics) from the Stockholm School of Economics in 2010. He also holds a PhD (agricultural economics) from Hokkaido University. Besides the **DCchoice** package described in Chapter 2 of this book, he is the author and the maintainer of the CRAN package called **ccgarch** that is designed for handling multivariate generalized autoregressive conditional heteroskedastic (GARCH) models.

Kazuo Sato is a professor at the College of Agriculture, Food and Environment, Rakuno Gakuen University, Japan. He received his PhD (agricultural economics) from Hokkaido University in 1999. His current research interests include analysis of consumers' preference for food safety and country-of-origin labeling of foods.

Chapter 1

Introduction

Abstract

Chapter 1 presents basic information on the stated preference (SP) methods—contingent valuation (CV), discrete choice experiments (DCEs), and best–worst scaling (BWS)—and outlines the role of this book. Section 1.1 explains the importance of R—free and open-source software—in the context of learning the SP methods. In Section 1.2, to facilitate entrance into the world of "empirical studies using SP methods," the three approaches adopted in the book, namely, focusing on the introductory-level variants of the three methods, making use of two or more example datasets to illustrate each method, and adding comments to the code, are explained. Thereafter, Section 1.3 outlines CV, DCEs, and BWS, limited to the variants used in this book. Section 1.4 introduces random utility theory and discrete choice models, which form the fundamental background for the three methods. Finally, Section 1.5 presents a summary of the rest of the book.

1.1 Stated preference methods and the role of R

Stated preference (SP) methods are a family of survey methods measuring people's preferences for alternatives based on decision-making in hypothetical choice situations. The family includes contingent valuation (CV), discrete choice experiments (DCEs), and best–worst scaling (BWS). These methods have been used to capture preferences for goods/services that are impossible or difficult to measure using traditional methods based on recorded market transactions, i.e., revealed preferences (RP), such as the environment and new products/services.

A common feature of these methods is that when applying them in these situations, they all require the design of hypothetical choice situations and analysis of choice behaviors. In general, application of an SP method can broadly be described by the following steps:

- Determine the goods/services to which the method will be applied.
- Design a (set of) hypothetical choice situation(s) according to the object and its hypothesized value domain.

- Ask the individuals in a sample drawn from the population, to indicate their preferences over alternatives in the hypothetical choice situation.
- Examine the factors affecting their choice behaviors, typically using econometric methods.
- Calculate various economic measures from the results.

The detailed content of each step varies according to the issue being targeted. As an example, let us consider a situation in which residents' benefits of an environmental conservation plan in a rural area must be measured. As the benefits include non-use values related to the environmental change, an SP method is an appropriate approach,[1] with the content thereof based on the following:

- Design a hypothetical choice situation in which residents in the rural area are asked to agree or disagree with an environmental conservation plan that also requires them to pay a certain amount to implement the plan.
- Ask the selected sample of the resident population to respond to the question posed in the hypothetical situation.
- Examine the effect of the specified amount of money on the answers.
- Quantify the benefit derived from the conservation plan.

As a second example, let us imagine a situation where a new public transport mode must be predicted for an area. Since the new public mode is intended to have new features that differ from those of existing public modes (bus, train), residents' preferences for the new public mode cannot be predicted accurately. Thus, an SP method is needed. The content of the steps followed is likely to be:

- Design hypothetical choice situations in which residents are requested to choose a mode from a set of transport modes (bus, train, car, and so on) including the new public transport mode, based on different expected travel costs and times.
- Select a sample of individuals from the resident population and ask them to select one mode in response to each question.
- Examine the relationship between the choices and the characteristics of the modes presented in the questions.
- Predict the share of each transport mode including the new one.

The alternatives from which respondents are required to choose may be goods and services, policy options, travel modes, and many more, and thus the method is widely applicable in a variety of disciplines. Furthermore, there ap-

[1]The total economic value of an environment includes both the use value and non-use value. The former is related to values generated from individuals' actual or possible use of the environment, while the latter is derived from the existence of the environment without the individuals' actual or possible use. While SP methods can estimate both use and non-use values, RP methods can estimate only the use value. Refer to Freeman III (1993) for details of the total economic value concept and methods for measuring values.

pear to be no restrictions on the type of targeted population, which is usually ordinary people such as citizens, residents, consumers, or travelers, although other types of agents are also targeted, like factory managers, product buyers, farmers, researchers, or medical doctors. Thus, it is clear that SP methods can be applied in many varied situations faced by people in their daily lives.

However, the wide range of applications of these methods could result in misunderstanding on the part of those who are not familiar with these methods, as they may be convinced that their way of using the methods is correct and that other ways do not exist or are incorrect. For example, people who are highly interested in environmental valuation may believe that the SP methods are limited to measuring economic values of the environment that cannot be captured by RP methods, such as the hedonic pricing method or travel cost method. People who engage in marketing research may believe that these methods are the only ones that can ascertain consumers' preferences for various product/service characteristics, which are useful in developing new products and services. Although it is true that these examples do indeed demonstrate the abilities of SP methods, the methods are by no means limited to these applications. Therefore, when learning SP methods for empirical applications, it is crucial to pay attention to both their common fundamental background and their wide-ranging applications.

Conversely, acquiring the broad knowledge related to the SP methods may be burdensome to some beginners wishing to apply the methods to their studies. Needless to say, excellent books about SP methods have been published (e.g., Mitchell and Carson, 1989; Louviere et al., 2000; Bateman et al., 2002; Hensher et al., 2005). Other books focus on experimental designs for SP methods (e.g., Street and Burgess, 2007; Raghavarao et al., 2011), or econometric analysis of discrete choice data (e.g., Ben-Akiva and Lerman, 1985; Train, 2009). These books certainly help in understanding the detailed theory and practice related to SP methods. However, the extensive detailed information therein may be excessive for beginners, thus preventing them from learning.

In addition to learning the theory and practice, beginners must understand how to use application software packages when designing choice situations (questions for the SP methods) and/or analyzing responses to questions. Several software packages or add-ins specializing in designing choice situations have been developed, while functions for discrete choice models are also provided in various statistical/econometric software packages. These software packages are very powerful and useful for empirical studies using the SP methods. Furthermore, such packages enable us to evaluate our understanding of the theory, which is particularly useful to those learning the theory without a teacher. However, these software packages are usually commercial products. It may be expensive for beginners to purchase a commercial software package for the purpose of only learning the SP methods. In addition, since the program code for such packages is compiled and thus not available in a human readable form, users are not able to read the source code to help them understand the theory and calculation processes better.

A way of overcoming these limitations is through the use of open-source software, where the source code is made available to the public under a particular license. A good example of such a package is R (R Core Team, 2014), which has the following advantages:

- The cost of using R is negligible if one has access to the Internet.
- The internal processes in R are open to the public; that is, the source code is available to the public under the Free Software Foundation's GNU General Public License.[2]
- R has a large collection of packages for statistics and graphics.
- R users can develop additional packages for R and make them available to the general public.
- There is an infrastructure for freely distributing additional R packages, called the Comprehensive R Archive Network (CRAN). More than 5,000 packages have been distributed on CRAN as of January 2014.[3]
- There is a central platform for the development of R packages called R-Forge, which is similar to CRAN, but hosts development versions of various R packages. Some of these are only available from R-Forge.
- Various books and documentation related to R written in both English and other languages have been published.
- There are active communities to assist with implementation issues.

Of course, R also has some disadvantages, although these are mostly offset by the advantages discussed above. The main limitation, especially for beginners, is that R was initially designed for use with a character user interface; that is, users are required to input commands for calculations and drawing graphics via a keyboard. However, a package called **Rcmdr** (Fox, 2005) attaches a graphical user interface (GUI) to R, enabling users to carry out their analyses by selecting appropriate pre-defined commands from the menu.[4] Therefore, by using R, beginners who are interested in the methods discussed in this book will be able to learn how to apply them with limited cost and psychological barriers.

1.2 Objective of this book

The objective of this book, aimed at beginners to these methods, is to demonstrate the implementation of SP methods using R. Although a number of these methods exist, only the core methods are discussed in this book: CV, DCEs, and BWS. The primary audience consists of undergraduate and graduate

[2]The license is given in full in the R software or can be obtained on the R website (http://www.r-project.org/).

[3]There are many mirror sites for CRAN. The reader should consult the CRAN web page on the R website when searching for a suitable mirror site located close by.

[4]Other GUIs have also been developed. Refer to the *Journal of Statistical Software*, volume 49, which is a special volume covering GUIs for R (http://www.jstatsoft.org/v49).

students in empirical economic fields, such as agricultural, environmental, transportation, and health economics, and in marketing research. This book would be suitable as a textbook for introductory-level hands-on courses or as supplemental reading in introductory-level courses related to these fields. Although it is envisaged that this book will be used primarily for teaching, novice researchers and practitioners in the three methods will also benefit from the contents thereof.

While writing this book, we kept in mind the idea of facilitating the reader's entrance into the world of "empirical studies using SP methods." Many excellent empirical studies have demonstrated the attractiveness of this area of research, and have drawn many users with an interest in these methods toward it. A range of excellent books exists to teach beginners the theory and practice of the SP methods. However, many of these texts deal with both introductory and advanced-level variants of the SP methods and thus assume a considerable degree of statistical and/or economic knowledge. We therefore focus on introductory-level variants of the three methods: single-bounded dichotomous choice (SBDC) and double-bounded dichotomous choice (DBDC) CV based on parametric and nonparametric approaches for analyzing responses; DCEs, based on orthogonal main-effect designs (OMEDs)[5] for designing choice situations and conditional logit (CL) and binary logit (BL) models for analyzing responses; and object case BWS, based on OMED or balanced incomplete block design (BIBD) for designing choice situations and counting and modeling approaches for analyzing responses.

Additionally, to ensure that beginners obtain valuable experience in the application of the three methods when executing actual code on their PCs, we focus on two aspects: using two or more example datasets to illustrate each method and include comments in the code.

Ways of implementing each method in R are described in terms of two or more example datasets. Examples in CV have been gathered from famous environmental valuation studies (e.g., the *Exxon Valdez* oil spill). DCE uses synthetic datasets of three examples related to food marketing and environmental valuation, while the two examples used to illustrate BWS refer to valuing environmental and food safety issues.

The examples are used to demonstrate the capabilities of the reviewed methods. However, the potential uses of these methods are not limited to the example situations. Therefore, we briefly present some contrived examples in previous empirical studies related to the three methods in Appendix B. Although this short appendix may not be of interest to advanced researchers/practitioners of these methods, we hope that it will be valuable to beginners wishing to use these methods in their empirical studies.

The source code for functions of the SP methods developed by the authors of this book is suitably commented. The comments are designed to help readers

[5]This book does not cover efficient designs for creating choice sets; refer to Louviere et al. (2000), Scarpa and Rose (2008), and Bliemer and Rose (2011) for information on this approach.

First question

The current water quality of lake A is such that the water is only suitable for agricultural irrigation. The proposed plan will improve the water quality to a level whereby people can enjoy swimming. This plan will require that your household pays $*X* annually. Do you agree or disagree with this plan?

1. I agree.
2. I disagree.

Second question (if the answer to the first question was "I agree")

If the amount of money that you would need to pay was $*BH*, would you now agree or disagree with this plan?

1. I agree.
2. I disagree.

Second question (if the answer to the first question was "I disagree")

If the amount of money that you would need to pay was $*BL*, would you now agree or disagree with this plan?

1. I agree.
2. I disagree.

Figure 1.1 *Example questions for CV.*

who are unfamiliar with the style of R coding to read and understand flows in each function.[6]

1.3 Overview of CV, DCEs, and BWS

1.3.1 Contingent valuation

CV is probably the most well known and most applied group of SP methods. Although there is a wide variety of CVs, we discuss only two variations in this book, namely, SBDC-CV (Bishop and Heberlein, 1979) and DBDC-CV (Carson, 1985; Hanemann, 1985). Of all the CV variations, these two are perhaps the most successful.

Figure 1.1 shows example questions for DBDC-CV that aims to measure environmental benefits of improving the water quality in a lake. The first question requests residents living in the basin of lake A to judge the environmental benefits of a plan to improve the quality of the water in the lake. In the CV question, two situations are explained to the respondents: the current water

[6] How to obtain the commented source code is explained in Chapter 5.

quality and the improved quality assuming the plan is implemented. Under these assumptions, the respondents are asked whether they agree with the plan and would be willing to pay a certain amount of money toward the implementation of the plan. The amounts (bids) that respondents are requested to contribute toward the plan are listed in advance (e.g., \$10, \$20, ...). Each respondent is requested to answer a question randomly aligned with one of the listed bids. An SBDC-CV survey requires respondents to answer only the first question in Figure 1.1, whereas a DBDC-CV survey requires them to answer both the first and second questions. The bid in the second question varies according to the response to the first question: a higher bid BH ($> X$) is displayed when the response to the first question is "I agree," whereas a lower bid BL ($< X$) is displayed when the response to the first question is "I disagree." The economic value of the environmental benefits can then be estimated by statistically examining the relationship between the bids presented in the questions and the responses to the questions (see Chapter 2 for the details).

1.3.2 Discrete choice experiments

DCEs were developed by Louviere and Woodworth (1983) and have since been improved and applied in various disciplines. The appeal of DCEs is that theoretically they have the ability to imitate almost any choice situation in our everyday lives. In general, respondents select one of two or more alternatives involving two or more attributes (one of which is a cost attribute). Although DCEs are also known as choice experiments, choice-based conjoint analysis, stated choice method, or attribute-based SP method, this book refers to them as DCEs as per Louviere et al. (2010). Of the many DCE variants, we focus on the most fundamental one, in which choice sets are created using OMED and responses are analyzed using a CL or BL model.

Figure 1.2 shows example questions for a DCE that aims to measure consumers' valuations of the attributes of apples. In general, choice tasks show two or more alternatives from which respondents are asked to select one. In this example, there are three alternatives: two apple alternatives and an opt-out option. The apple alternatives are constructed from five attributes: variety, country of origin, cultivation method, production information, and price per piece. For example, Apple 1 in Q1 is of the Red Delicious variety, produced using a standard cultivation method in the United States, sold for \$2.2 per piece, and its production information is accessible on the Internet. The combinations of attribute levels are usually determined during the design of the experiments. This book uses OMEDs to create the combinations: an OMED satisfying the conditions of the attributes and attribute levels is selected and then the combinations are created using the OMED (see Chapter 3 for ways of designing choice sets using OMEDs). Respondents evaluate each apple based on these attribute levels, and then select their preferred option from the three alternatives. In general, a choice task is repeated two or more times with each

Q1 Which apple would you like to purchase?

	Apple1	Apple2
Variety	Red Delicious	Royal Gale
Country of origin	United States	New Zealand
Cultivation method	Standard	Organic
Production information	Available	Unavailable
Price per piece	$2.2	$3.0

1. I would like to purchase apple 1.
2. I would like to purchase apple 2.
3. I would not like to purchase either of these.

...

Q8 Which apple would you like to purchase?

	Apple1	Apple2
Variety	Fuji	Golden Delicious
Country of origin	Japan	United States
Cultivation method	Low chemicals	Organic
Production information	Unavailable	Unavailable
Price per piece	$2.4	$2.8

1. I would like to purchase apple 1.
2. I would like to purchase apple 2.
3. I would not like to purchase either of these.

Figure 1.2 *Example questions for a DCE.*

task containing different alternatives; i.e., the question is repeated eight times in this example, with each apple's characteristics differing in each question. Consumers' preferences for each attribute level are measured by statistically analyzing the relationship between the alternative characteristics and the respondents' choice behavior (see Chapter 3 for further details).

1.3.3 Best–worst scaling

BWS was developed in the late 1980s to expand the capability of DCEs (Flynn, 2010). The development of the theoretical foundations thereof is ongoing (Marley and Louviere, 2005; Marley et al., 2008; Marley and Pihlens, 2012). Although it has three variations, namely, object case BWS, profile case BWS, and multi-profile case BWS, the discussion in this book is limited to object case BWS, as this is the most fundamental variation of the three.

Figure 1.3 shows some example questions for object case BWS. In BWS, after listing a set of candidate items, subsets are created according to the design of the experiment, such as OMED or BIBD, and each subset is converted into

Q1 Please select the city in which you would most like to live (most attractive) and the city in which you would least like to live (least attractive).

Most attractive	City	Least attractive
	Tokyo	
	Kyoto	
	Okayama	
	Naha	

...

Q13 Please select the city in which you would most like to live (most attractive) and the city in which you would least like to live (least attractive).

Most attractive	City	Least attractive
	Sapporo	
	Sendai	
	Kyoto	
	Naha	

Figure 1.3 *Example questions for BWS.*

a question. Then, respondents select the best and worst items in each question. In this example, respondents are required to evaluate different cities in Japan from the point of view of their desire to live there. Q1 presents four cities: Tokyo, Kyoto, Okayama, and Naha. These cities are selected from a candidate list of cities (which is omitted here), and respondents select two cities—one where they would most like to live, and the other where they would least like to live. In general, questions for BWS are repeated several times, with each question providing different combinations of items (13 BWS questions are used in this example). The relative importance of items is measured by analyzing the relationship between the combination of items and the responses based on the counting approach (i.e., the numbers of times each item is selected as the best and the worst, respectively) or modeling approach (i.e., discrete choice model analysis). Previous studies frequently examined the differences in the measure between two or more countries (i.e., international comparison) or segmented respondents using individual measurement levels (i.e., market segmentation) (see Chapter 4 for the details of these approaches).

1.4 Random utility theory and discrete choice models

The variants of CV, DCEs, and BWS that are the focus of this book can be analyzed using discrete choice models derived from the common theoretical background of random utility theory. This section provides a fundamental explanation of random utility theory and discrete choice models common to CV, DCEs, and BWS.[7] The reader is referred to Chapters 2, 3, and 4 for a detailed explanation of respondents' decision-making with regard to questions for CV, DCEs, and BWS, respectively, from the viewpoint of random utility theory.

1.4.1 Random utility theory

Let there be two or more alternatives (including a possible opt-out option, such as "I select none of these") in a choice set (S) for individual n in a choice situation. For simplicity, we assume that the choice set is common to all individuals. Further assume that individual n gains utility U_{in} from alternative i:

$$U_{in} = V_{in} + e_{in}, \tag{1.1}$$

where V_{in} is individual n's systematic (representative) component of the utility of alternative i, while e_{in} is individual n's random component of the utility of alternative i. The former is determined by the attribute variables related to alternative i, while the latter cannot be observed, and is thus, random to the analyst.

Assume that individual n selects alternative i in S if and only if it provides the greatest utility of the available alternatives. That is, for $\forall i \neq j$, we have

$$U_{in} > U_{jn} \tag{1.2}$$

is equivalent to

$$V_{in} - V_{jn} > e_{jn} - e_{in}. \tag{1.3}$$

For example, a respondent asked to answer the first question for CV in Figure 1.1 is faced with a choice set of two alternatives, that is, "I agree with the plan that requires an annual payment of $X" in return for improved water quality compared with the current quality, and "I disagree with the plan." If the respondent selects "I agree," we can interpret this to mean that the respondent's utility of this alternative is greater than that of the other. In the case of the DCE in Figure 1.2, there are three alternatives in the choice set, that is, two apple alternatives and an opt-out option. If a respondent selects "I would not like to purchase either of these" in Q1, we can interpret this to mean that the respondent gains the most utility from the opt-out option. In

[7] See Ben-Akiva and Lerman (1985) and Train (2009) for details of random utility theory and discrete choice models.

the same way, if a respondent selects Kyoto as the most attractive city and Tokyo as the least attractive in Q1 in Figure 1.3, we can interpret this to mean that Kyoto and Tokyo provide the maximum and the minimum utility, respectively, for the respondent.

Now, since the analyst cannot observe the right-hand side of Eq. (1.3), we assume that s/he can only calculate the probability that the left-hand side of the equation is greater than the right-hand side. Accordingly, the probability that individual n will select alternative i from choice set S is:

$$\begin{aligned} P_n(i) &= \Pr(U_{in} > U_{jn}) \\ &= \Pr(V_{in} - V_{jn} > e_{jn} - e_{in}), \quad \forall i \neq j. \end{aligned} \tag{1.4}$$

1.4.2 Discrete choice models

We can derive a number of discrete choice models by assuming specific conditions for Eq. (1.4). For example, assuming that there are three or more alternatives in S and that e_{in} are independent and identically distributed Gumbel variables, the following CL model is derived:

$$P_n(i) = \frac{\exp(V_{in})}{\sum_{j \in S} \exp(V_{jn})}. \tag{1.5}$$

This model is a fundamental one used in DCEs and BWS.[8]

Furthermore, a BL model, which is the fundamental model used in SBDC-CV, is obtained if there are only two alternatives in Eq. (1.5) and given that the assumption for e_{in} holds:

$$\begin{aligned} P_n(i) &= \frac{\exp(V_{in})}{\exp(V_{in}) + \exp(V_{jn})}, \\ &= \frac{1}{1 + \exp(V_{jn} - V_{in})}. \end{aligned} \tag{1.6}$$

The systematic component of utility is assumed to be a linear additive function of the independent variables X_{ikn}:

$$V_{in} = \beta_0 + \sum_{k=1}^{K-1} \beta_k X_{ikn}, \tag{1.7}$$

where β_0 is a constant, β_k is the coefficient of variable X_{ikn}, and K is the number of coefficients including the constant.

[8]It is noted that the multinomial logit (ML) model is similar to the CL model. By definition, the systematic component of utility in the ML model depends on variables that are specific to the individual (e.g., gender, age, and income), whereas in the CL model, it depends on variables specific to the given alternative. Practically, the systematic component of utility is frequently modeled to be dependent on both individual- and alternative-specific variables. Therefore, the model names are frequently confused. See Greene (2002), Wooldridge (2002), or Cameron and Trivedi (2005) for details of each model.

To estimate the unknown parameters, we can apply the maximum likelihood technique. For a given sample of N independent observations, the log-likelihood function is written as:

$$\ln L = \sum_{n=1}^{N} \sum_{i \in S} d_{in} \ln P_n(i), \tag{1.8}$$

where d_{in} is an indicator variable equal to 1 if individual n selects alternative i, and 0 otherwise. The parameter estimates are obtained by maximizing the log-likelihood function.

After estimating the parameters, it is standard practice to conduct statistical tests related to the parameters to evaluate the estimated model. The most basic test is whether each individual coefficient in the model is statistically different from the value 0 (z-test). The test statistic is calculated by dividing an estimate by its standard error ($s.e.$). The resulting quantity, often referred to as the "z-value" or "asymptotic t-value," is asymptotically distributed as a standard normal distribution under the null hypothesis. The formula for obtaining the z-value is given by

$$z = \frac{\hat{\beta}_k}{s.e.(\hat{\beta}_k)}. \tag{1.9}$$

The null of $\beta_k = 0$ is rejected at the $100\alpha\%$ level of significance if z is greater than the $\alpha/2$-percentile (conventionally $\alpha = 0.05$ for the two-sided test) of a standard normal distribution.

To conduct a test for multiple restrictions on the parameters, the likelihood ratio test can be used. The test statistic LR is defined as:

$$LR = -2(\ln L_R - \ln L_U), \tag{1.10}$$

where $\ln L_R$ and $\ln L_U$ are the log-likelihood values at convergence for a restricted and unrestricted model, respectively. The null hypothesis in an LR test is that the restrictions are true. Under the null, the LR test statistic is asymptotically distributed as a χ^2 distribution with D degrees of freedom, which is equal to the number of restrictions. If the test statistic is greater than the percentiles of the $\chi^2(D)$ distribution for an α level of significance, the null hypothesis (the restricted model) is rejected. An example of the test is whether two coefficients are equal. In this case, the restricted model is that one coefficient, β_1, is equal to the other, β_2, whereas in the unrestricted model there is no restriction on the coefficients. In this case, the number of restrictions is 1. Another example is to test whether all coefficients are equal to 0. In this case, the restricted model is that all coefficients are restricted to zero, whereas the unrestricted model has no restrictions. In this example, the number of restrictions is the number of coefficients restricted to zero. It should be mentioned that the LR test is only applicable in cases where the restricted model is a special case of the unrestricted model. Such a special case is obtained with restrictions on the parameters to be tested.

When we select a model from a set of non-nested or nested model variants, the Akaike Information Criterion (AIC), or Bayesian Information Criterion (BIC) is often used, as defined below:

$$AIC = -2\ln L_{\hat{\beta}} + 2K, \tag{1.11}$$

$$BIC = -2\ln L_{\hat{\beta}} + (\ln N)K. \tag{1.12}$$

For each criterion, a lower value is preferable.

It is often useful to compute goodness-of-fit measures. A well-known measure is ρ^2 (rho-squared)—also called McFadden's R^2 or pseudo R^2—which is defined as:

$$\rho^2 = 1 - \frac{\ln L_{\hat{\beta}}}{\ln L_0}, \tag{1.13}$$

where $\ln L_{\hat{\beta}}$ and $\ln L_0$ are, respectively, the log-likelihood values for the estimated model and for the model without coefficients (i.e., all coefficients are equal to zero).[9] The measure adjusted by the number of coefficients (K) is defined as:

$$\bar{\rho}^2 = 1 - \frac{\ln L_{\hat{\beta}} - K}{\ln L_0}. \tag{1.14}$$

It should be noted that ρ^2 is not comparable with R^2 in a linear regression model. Since ρ^2 does not produce high values, a value between 0.2 and 0.4 reflects a good model fit (Louviere et al., 2000).

Sometimes it is hard to define a null model as the base of ρ^2. For instance, in the DBDC-CV analysis in Chapter 2, setting all the parameters except the intercept induces ln(0) in the log-likelihood function.

We can extract valuable measures from an estimated model, one of which is the marginal/willingness-to-pay (M/WTP). (M)WTP is the amount of money that respondents are willing to pay to obtain a further (additional) non-monetary attribute. As an example, we consider the SBDC-CV in Figure 1.1. Under Eq. (1.6), the systematic component of the utility function is assumed to be:

$$V_{jn} - V_{in} = \beta_0 - \beta_1 X, \tag{1.15}$$

where X is the amount of money that the respondents are asked to pay in the question. After estimating the model, the median WTP is calculated as:

$$\frac{\hat{\beta}_0}{\hat{\beta}_1}. \tag{1.16}$$

As another example, consider the DCE in Figure 1.2. It is assumed that the responses are analyzed using Eqs. (1.5) and (1.7). Here, we focus on the MWTP

[9] There is another version of ρ^2, which uses the log-likelihood value of the estimated model with only constant terms instead of $\ln L_0$.

for availability of production information: variable X_I equal to 1 if it is available, and 0 otherwise. By using the estimated coefficients $\hat{\beta}_I$ and $\hat{\beta}_P$ corresponding to variables X_I and X_P (price attribute variable), respectively, the MWTP is calculated as:

$$-\frac{\hat{\beta}_I}{\hat{\beta}_P}. \tag{1.17}$$

When constructing confidence intervals for these measures, three methods are widely used in practice: the delta, Krinsky and Robb (parametric bootstrap), and (nonparametric) bootstrap methods.[10]

The first method calculates confidence intervals based on estimates. For example, to find the $100(1-\alpha)\%$ confidence interval for MWTP in the case of the DCE mentioned above, the delta method computes:

$$\frac{\hat{\beta}_I}{\hat{\beta}_P} \pm z_{\alpha/2}\sqrt{\left(\frac{\hat{\beta}_I}{\hat{\beta}_P}\right)^2 \cdot \left(\frac{\text{var}(\hat{\beta}_I)}{\hat{\beta}_I^2} + \frac{\text{var}(\hat{\beta}_P)}{\hat{\beta}_P^2} - \frac{2\text{cov}(\hat{\beta}_I, \hat{\beta}_P)}{\hat{\beta}_I \hat{\beta}_P}\right)}, \tag{1.18}$$

where $z_{\alpha/2}$ is the $\alpha/2$ percentile of a standard normal distribution.

The other two methods calculate confidence intervals using simulation techniques with different settings. In the Krinsky and Robb method (Krinsky and Robb, 1986, 1990), a coefficient vector is replicated R times from a multivariate normal distribution with a vector of estimated coefficients and a variance-covariance matrix of estimated coefficients. Then, the measures are computed for each replicated coefficient vector. As a result, we are able to build an empirical distribution of each of the measures, thereby obtaining the associated confidence intervals. For example, the lower and upper bounds of the 95% confidence interval correspond to the 0.025 and 0.975 percentiles of the measures, respectively. The bootstrap method resamples the data at hand and repeatedly estimates the model with the bootstrapped data to formulate an empirical distribution of the associated measure. The upper and lower bounds of the interval are determined as in the Krinsky and Robb method.

It is often necessary to calculate the difference between two measures that are independently estimated: for example, a comparison of WTP to improve the quality of water in lakes A and B. Poe et al. (2005) proposed a complete combinatorial method for calculating the difference between two independent empirical distributions and conducting a statistical test related to the difference. Let us assume that X and Y are independent empirical distributions depicted by vectors $\mathbf{x} = (x_1, x_2, ..., x_m)$ and $\mathbf{y} = (y_1, y_2, ..., y_n)$, respectively. The null hypothesis is $X - Y = 0$, while the alternative hypothesis is $X - Y > 0$. The complete combinatorial method provides a one-sided significance level of

[10] Although each method has both advantages and disadvantages as pointed out in the literature, we do not discuss the adequacy of the methods. For a comparison of the methods, the interested reader is referred to Hole (2007) and the references therein.

Table 1.1 *Main packages used in this book*

Package	Author	Title
DCchoice	Nakatani (2014)	Analyzing dichotomous choice contingent valuation data
Ecdat	Croissant (2014)	Datasets for econometrics
support.CEs	Aizaki (2014b)	Basic functions for supporting an implementation of choice experiments
survival[a]	Therneau (2014)	Survival analysis, including penalized likelihood
stats[a]	R Core Team (2014)	The R stats package
mded	Aizaki (2012b)	Measuring the difference between two empirical distributions
DoE.base	Grömping (2014b)	Full factorials, orthogonal arrays and base utilities for DoE packages
crossdes	Sailer (2013)	Design and randomization in crossover studies
support.BWS	Aizaki (2014a)	Basic functions for supporting an implementation of best–worst scaling

[a] denotes a basic/recommended R package, details of which are given in Chapter 5.

the null hypothesis, which is calculated by dividing the number of cases where $x_i - y_j \leq 0$ in all combinations of i and j by $m \times n$. For example, if the significance level is 0.034, the null hypothesis is rejected at the 5% level. Since this approach needs two empirical distributions, the Krinsky and Robb or bootstrap method must be applied in advance.

1.5 Summary of the rest of this book

This book consists of five chapters, including this chapter, and two appendices. The core content is contained in Chapter 2 (CV), Chapter 3 (DCEs), and Chapter 4 (BWS). Fundamentals of R are explained in Chapter 5, while the appendices describe other R packages related to this book and examples of applications using CV, DCEs, and BWS in previous empirical studies in a variety of fields. The rest of this book is summarized below (see also Table 1.1 for the main packages used in this book).

Chapter 2 deals with the implementation of SBDC- and DBDC-CV in R. After outlining the two types of CV, the theoretical foundations of parametric and nonparametric estimation methods for mean and median WTPs are presented in Section 2.2. Although there are variations in each of the parametric and nonparametric approaches, this book focuses on the utility difference approach (Hanemann, 1984; Hanemann et al., 1991) as a parametric estimation method, and the approaches of Kriström (Kriström, 1990) and Kaplan–Meier–Turnbull (Carson and Hanemann, 2005) as nonparametric estimation methods. Section 2.3 introduces R functions, contained in the **DCchoice** package (Nakatani, 2014), for implementing these methods. After presenting an overview of the package, we explain how to install and load the package and prepare datasets suitable for use by the package functions. Since the package depends indirectly on a package available from the Bioconductor website (http://www.bioconductor.org/), the way the package is installed into R from the website is also explained. In addition, this section briefly introduces four datasets used as examples in the next two sections. Sections 2.4 and 2.5 discuss the basic usage of functions included in **DCchoice** and present example code using the functions. Section 2.4 focuses on parametric estimations of WTP and also explains how to construct confidence intervals with respect to the WTP estimates using the Krinsky and Robb and bootstrap methods. Similarly, Section 2.5 explains nonparametric estimation of WTPs.

Chapter 3 deals with the implementation of DCEs in R. Section 3.2 defines the basic terms of DCEs, such as attributes, attribute levels, alternatives, and choice sets, as well as categories of DCEs. For example, DCEs can be categorized according to the format of the choice sets, yielding categories such as binary DCEs and (general) DCEs, or labeled DCEs and unlabeled DCEs. This section also outlines the three steps for implementing DCEs: examining a hypothetical choice situation and creating choice sets; preparing and conducting a survey; and preparing a dataset and conducting the analysis. Section 3.3 introduces the R functions implementing these methods. Section 3.4 introduces three examples on how to use these R functions. The first example illustrates consumers' evaluations of the characteristics of rice based on an unlabeled DCE. The second example investigates consumers' evaluations of three types of pork based on a labeled DCE. The third example involves conducting a binary DCE in which residents in two regions are requested to evaluate a rural environmental conservation plan.

Chapter 4 discusses the implementation of object case BWS in R. Section 4.2 explains how to construct choice sets for BWS, including an approach using a two-level OMED and one using a BIBD, as well as how to analyze responses to the BWS questions, that is, an approach counting the numbers of times each item is selected as the best and the worst, and one applying a conditional logit model. Section 4.3 explains how to implement these approaches in R. Section 4.4 demonstrates two hypothetical examples using these R functions. The first example using a two-level OMED elicits citizens' preferences regarding the multifunctionality of agriculture such as forming the landscape or preserving biodiversity. The second example using a BIBD explores consumers' preferences over seven fruits. This chapter includes an appendix briefly explaining case 2 (profile case) BWS and case 3 (multiprofile case) BWS.

Chapter 5 focuses on basic operations implemented in R, including installing R and its various packages, reading the code of a function, loading data from external files, handling missing observations, manipulating vectors and matrices, creating dummy variables, and drawing graphs. References are provided for further reading.

Appendix A gives additional information on R packages related to the topics covered in the previous chapters, namely, CV, discrete choice models, cluster analysis, principal component analysis, factor analysis, and conjoint analysis. Appendix B provides concise examples of the intricacies of using CV, DCEs, and BWS in previous empirical studies. The examples are divided into four main topics: providing information to respondents; using product/service samples; cost–benefit analysis and deliberative monetary valuation; and using SP study results in simulations.

Chapter 2

Contingent Valuation

Abstract

Chapter 2 explains the implementation of single- and double-bounded dichotomous choice contingent valuation in R. After outlining the two types of contingent valuation, the theoretical foundations of parametric and nonparametric estimation methods for mean and median willingness-to-pays (WTPs) are presented in Section 2.2. This chapter focuses on the utility difference approach as a parametric estimation method, and the non-parametric approaches of Kriström and Kaplan–Meier–Turnbull. Section 2.3 introduces the R functions, contained in the **DCchoice** package, for implementing these methods. Since this package depends indirectly on a package available from the Bioconductor website, the way the package is installed into R from the website is also explained. Furthermore, this section briefly introduces four datasets used as examples in subsequent sections. Sections 2.4 and 2.5 discuss the basic usage of the functions included in **DCchoice** and present example code using these functions. The two sections focus on parametric and nonparametric estimation of WTPs, respectively.

2.1 Introduction

Contingent valuation (CV) is a survey-based methodology for eliciting the values people place on goods, services, and amenities (Boyle, 2003).[1] Although CV was used mainly in the field of environmental economics in the early years of its existence, today it has a broad range of applications in several areas, for example:

Medical and health care research Health economics is one of the few areas in which surveys have been used for the purpose of evaluating economic policies since the early years of CV development (Carson and Hanemann, 2005), and there are a wide variety of applications of CV in this field:

- WTP/WTA for treatment of illnesses (Zethraeus, 1998; Dranitsaris et al., 2000; Haefeli et al., 2008)
- WTP for drugs (O'Brien et al., 1995; Lang, 2005, 2010)

[1]For a detailed history of the CV method, see Mitchell and Carson (1989), Smith (2006), and Carson (2012).

- WTP for vaccines (Whittington et al., 2002; Kim et al., 2008; Lieu et al., 2008)
- WTP for health insurance (Dong et al., 2003; Ying et al., 2007; Gustafsson-Wright et al., 2009)
- Disutility of queues for surgery (Bishai and Lang, 2000)
- WTP/WTA for informal care (van den Berg et al., 2005; de Meijer et al., 2010)
- Quality improvements in public healthcare facilities (Weaver et al., 1996)

Food research In the food research field, the use of CV surveys includes assessing WTP for food products that may be introduced into the market and new food products only available to a small subset of consumers. For example:

- Genetically modified food (Li et al., 2002; Grimsrud et al., 2004; Lusk, 2003; Curtis and Moeltner, 2006)
- Low-pesticide or pesticide-free food (Fu et al., 1999; Boccaletti and Nardella, 2000; Cranfield and Magnusson, 2003)
- Eco-friendly food (Loureiro et al., 2002; Loureiro, 2003; Kimura et al., 2010)
- Food labeling (Lusk and Fox, 2002; Loureiro et al., 2006; Tonsor and Shupp, 2009; Tonsor and Wolf, 2011)

Culture In the field of cultural economics, values of cultural facilities, cultural activities, and historic ruins that include "non-use value," have been estimated by CV. For example:

- Theatre and museums (Martin, 1994; Hansen, 1997; Sanz et al., 2003; Bedate et al., 2009)
- Public libraries (Harless and Allen, 1999; Aabø, 2005; Delaney and O'Toole, 2006)
- Performing arts companies (Evans, 1999)
- Public cultural programs (Santagata and Signorello, 2000)
- Cultural heritage conservation (Lockwood et al., 1996; Pollicino and Maddison, 2001; Whitehead and Finney, 2003; del Saz Salazar and Marques, 2005)

Risk CV has frequently been used in the context of risk analysis to measure the value of risk reduction in various situations and to estimate the value of statistical life (VSL). For example:

- Job-related fatal accidents (Gerking et al., 1988)
- Transportation (McDaniels, 1992; Persson et al., 2001; Carlsson et al., 2004; Hultkrantz et al., 2006; Bhattacharya et al., 2007)
- Crime (Ludwig and Cook, 2001; Atkinson et al., 2005)
- Reduction in mortality (Krupnick et al., 2002)

There are many "elicitation formats" in CV. In the early years of CV, the bidding game and open-ended format were widely used. In the bidding game, the respondent's WTP is elicited through an auction-like procedure, whereas in the open-ended format, respondents are directly asked what their willingness-to-pay (WTP) is; e.g., "What is the most you would be willing to pay for $\cdots$?" With these formats, "the survey responses yield a direct measure of WTP which requires little or no further analysis" (Hanemann and Kanninen, 1999). After the epoch-making work of Bishop and Heberlein (1979), however, the single-bounded dichotomous choice (SBDC) format, and later, the double-bounded dichotomous choice (DBDC) format became popular. Unlike the earlier formats, binary or interval data are associated with these formats. To obtain a WTP value from these responses, some kind of statistical model is needed. The accompanying package **DCchoice** (Nakatani, 2014) provides parametric and nonparametric estimation of SBDC and DBDC CV data.

The chapter is organized as follows. Following the introduction, Section 2.2 explains the basic theory of SBDC-CV and DBDC-CV. Section 2.3 introduces the R functions, contained in the **DCchoice** package, for the SBDC- and DBDC-CV methods, the usage of which is discussed in Sections 2.4 and 2.5, respectively. Example code and datasets are also presented.

2.2 Overview of contingent valuation

In this section, the basic theoretical background of SBDC-CV and DBDC-CV is presented mainly based on the work by Bateman et al. (2002), Carson and Hanemann (2005), and Hanemann and Kanninen (1999). The description given here is largely limited to the methods available in the **DCchoice** package. For a more comprehensive economic and econometric theory of CV, please refer to the above literature.

2.2.1 Elicitation formats

Single-bounded dichotomous choice

The SBDC format (close-ended, referendum methods, take-it-or-leave-it approach) was first proposed by Bishop and Heberlein (1979). In the previous elicitation format, researchers directly asked respondents the value of their maximum WTP. For example, using an open-ended question, respondents were asked something like "What is the most you would be willing to pay for the change?" Although this question is very straightforward, it is not easy for the respondents to answer, particularly if they are unfamiliar with the pricing.

In SBDC format, respondents are asked something like "If this change cost you $\$A$, would you be willing to pay this amount?" This question is easier for respondents to answer because they merely have to exercise their judgment in deciding whether the price is acceptable, in the same way as making decisions about shopping, for example. Respondents only need to answer "yes" or "no" in SBDC format, according to whether their WTP is higher or lower than the bid presented to them. Thus, binary data are obtained using this format.

Double-bounded dichotomous choice

Although SBDC is easy for respondents to answer, "it is statistically less efficient and requires a larger sample to attain a given level of precision" (Hanemann et al., 1991). Thus, DBDC (dichotomous choice with follow-up, double referendum) was proposed by Hanemann (1985) and Carson (1985) to improve the efficiency of SBDC.

In the DBDC format, respondents are asked a second question immediately after answering the first SBDC style question. Usually, the bid included in the second question is somewhat higher than the initial bid if the respondent answered "yes" to the first question, and somewhat lower if the respondent answered "no" to the first question. With this second ("follow-up") question, more information about the respondent's WTP is obtained. Hanemann et al. (1991) showed that the DBDC approach is asymptotically more efficient than the SBDC approach.

For example, assume that the initial question is "If this change costs you \$$A$, would you be willing to pay this amount?" If the respondent answered "yes" to this initial question, the follow-up question might be "What if it cost you \$$BH$ instead of \$$A$ to obtain this change, would you be willing to pay this increased amount?" The second bid BH is higher than the initial bid A. The follow-up question for those answering "no" to the initial question might be "What if it cost you \$$BL$ instead of \$$A$ to obtain this change, would you be willing to pay this lesser amount?" In this case, the second bid BL is lower than the initial bid A.

In the DBDC question, there are four possible response outcomes: (yes, yes); (yes, no); (no, yes), and (no, no). If respondent n's answer is (yes, yes), the researcher can infer that $WTP_n > BH$ (where WTP_n is the WTP of respondent n). Similarly, (yes, no) implies that $A < WTP_n < BH$, (no, yes) implies that $BL < WTP_n < A$, and (no, no) implies that $WTP_n < BL$. Therefore, the data from DBDC are interval data.[2] Some kind of statistical model is needed to obtain point WTP estimates from these data.

2.2.2 Bid design

Many papers have been written about the bid design for SBDC and DBDC (e.g., Boyle et al., 1988; Duffield and Patterson, 1991; Cooper and Loomis, 1992; Kanninen, 1993a,b; Alberini, 1995). Until the 1980s, the commonly used bid design strategy for SBDC was the one employed by Bishop and Heberlein (1979); that is, determining the highest bid by some prior information, with the other bids chosen roughly according to log-linear intervals (Duffield and Patterson, 1991). In those early days, the goal of an optimal sampling procedure was to estimate the entire WTP distribution as accurately as possible.

In the 1990s, papers suggesting the use of an optimal experimental design for CV surveys began to appear. These papers set a more accurate estimation

[2] A binary variable is a special case of interval data.

of parameters or WTP as the goal of the experimental research. Two design criteria are frequently used: *d*-optimality and *c*-optimality. Given the total size of the sample, d-optimality seeks to minimize the asymptotic variances of the estimators by maximizing the determinant of the Fisher information matrix, whereas c-optimality minimizes the confidence interval of the estimated WTP.

However, to calculate these optimal design criteria, values of the parameters are needed. This is a paradoxical situation because if the true values of the parameters were available, one would not have to conduct a CV survey. Therefore, instead of the true values of the parameters, some kind of prior information about the parameters is used to calculate the optimal design criteria. However, if the prior information is not reliable, the corresponding design would not be reliable either. One of the best possible ways is to use the sequential design procedure proposed by Kanninen (1993a). It is a commonly accepted practice to apply information from a pretest, which is often conducted by open-ended CV or payment card, to construct the bid design for SBDC and DBDC.

2.2.3 Parametric estimation method

Parametric estimation of SBDC data

There are two major parametric estimation approaches for analyzing SBDC data. In the "utility difference approach" (Hanemann, 1984), the specification for the bid function is derived from an explicit formulation of the indirect utility function. On the other hand, in the "bid function approach" (Cameron and James, 1987; Cameron, 1988), a model of the bid function is built directly. We adopt the utility difference approach here, because it is more explicitly consistent with neoclassical economic theory.

Now we consider the case in which a researcher would like to value a change in the provision of a non-market good from its present level q_0 to a higher level q_1. We assume that the change from q_0 to q_1 increases the level of utility of an individual.

Assume that we can write the true indirect utility function $U(q, y)$, where q is the level of provision of the non-market good and y is the individual's income. For simplicity, we suppress all the other arguments like the prices of the market goods and the attributes of the individual.

In most cases using CV surveys, the value of the change in the non-market good is defined by the Hicksian compensating variation, C, which satisfies

$$U(q_1, y - C) = U(q_0, y). \tag{2.1}$$

C can be interpreted as the maximum WTP.

The researcher can ask respondents directly the value of C using an open-ended question like "What is the most you would be willing to pay for the change?" However, as mentioned before, this type of question is not easy for the respondents to answer. Instead, in SBDC-CV the researcher asks the respondents a question like "If it cost you $t to obtain this change, would you be willing to pay this amount?" Each respondent, who is supposed to know his/her own true indirect utility function, would answer "yes," if

$$U(q_1, y - t) \geq U(q_0, y) \tag{2.2}$$

and "no," otherwise.

Things are a little more complicated for the researcher, however, because a respondent's true indirect utility function is unknown to him/her. So, from the researcher's viewpoint, a respondent's answer has some stochastic aspect.

According to random utility theory as explained in Chapter 1, an individual's indirect utility function consists of a systematic component $V(q, y)$ and a random component e:

$$U(q, y) = V(q, y) + e. \tag{2.3}$$

In this context, an individual is assumed to answer "yes" to a SBDC-CV question with bid t if

$$V(q_1, y - t) + e_1 \geq V(q_0, y) + e_0$$
$$V(q_1, y - t) - V(q_0, y) \geq e_0 - e_1.$$

This expression is equivalent to Eq. (1.2) in Chapter 1.

The indirect utility function can take various forms. The simplest form is the linear utility model, given by

$$V_j = \alpha_j + \beta y + e_j \quad j = 0, 1, \tag{2.4}$$

where β can be interpreted as the marginal utility of income.

In this model, an individual's maximum WTP (denoted as C in the formulation below) satisfies

$$\alpha_0 + \beta y + e_0 = \alpha_1 + \beta(y - C) + e_1 \tag{2.5}$$

and therefore, the corresponding bid function is given by

$$C = \frac{\alpha + e}{\beta}, \tag{2.6}$$

where $\alpha = \alpha_1 - \alpha_0$ and $e = e_1 - e_0$.

In the linear utility model, a respondent will answer "yes" to a SVDC-CV question with bid t, if

$$V_1 - V_0 = \alpha_1 + \beta(y - t) + e_1 - (\alpha_0 + \beta y + e_0) = \alpha - \beta t + e \geq 0. \tag{2.7}$$

Note that $\alpha - \beta t + e$ in (2.7) is the utility difference. V_j in itself cannot be observed. If e is a standard normal random variable and the probability of obtaining a "yes" response to bid t is $P^y(t)$, the response formula becomes

$$P^y(t) = 1 - \Phi(-\alpha + \beta t) = \Phi(\alpha - \beta t), \tag{2.8}$$

which is called the probit model. If e is a standard logistic random variable,

$$P^y(t) = \frac{1}{\exp(-\alpha + \beta t)}, \tag{2.9}$$

which is called the logit model. If the Weibull distribution with $\ln(t)$ is assumed, the probability of obtaining a "yes" response to the bid t is given by

$$P^y(t) = \exp\{-\exp(-\alpha + \beta \ln(t))\}. \tag{2.10}$$

The normal and logistic specifications are particularly popular because of their relative simplicity (Bateman et al., 2002). But, if the researcher believes that the WTP cannot be negative for some reason, alternative choices are the log-normal and log-logistic distributions. By replacing t in Eq. (2.8) with $\ln(t)$, we obtain the log-normal distribution model:

$$P^y(t) = 1 - \Phi(-\alpha + \beta \ln(t)) = \Phi(\alpha - \beta \ln(t)). \tag{2.11}$$

Moreover, replacing t in Eq. (2.9) with $\ln(t)$ yields the log-logistic model, which was used in Bishop and Heberlein (1979).

$$P^y(t) = \frac{1}{\exp(-\alpha + \beta \ln(t))} \tag{2.12}$$

For a given sample of n independent observations, the log-likelihood function can be written as Eq. (1.7) in Chapter 1. For example, if e is a standard normal random variable, and thus $P^y(t) = \Phi(\alpha - \beta t)$, the corresponding log-likelihood function is given by

$$\ln L = \sum_{n=1}^{N} \left[d_n \ln \{\Phi(\alpha - \beta t_n)\} + (1 - d_n) \ln \{1 - \Phi(\alpha - \beta t_n)\} \right], \tag{2.13}$$

where d_n is an indicator variable that takes the value one if respondent n answers "yes" to the offered bid t_n and zero if they answer "no." In the case of logit model, the corresponding log-likelihood function is given by

$$\ln L = \sum_{n=1}^{N} \left[d_n \ln \left\{ \frac{1}{\exp(-\alpha + \beta t_n)} \right\} + (1 - d_n) \ln \left\{ 1 - \frac{1}{\exp(-\alpha + \beta t_n)} \right\} \right]. \tag{2.14}$$

If the researcher is interested in not only the value of WTP, but also what characteristics of the individual influence the utility change, α can be specified as a function of the individual's characteristics. In most cases, a simple linear form is employed, such as

$$\alpha = \gamma + \sum_{k=1}^{K-1} \gamma_k X_k, \tag{2.15}$$

where X_k, $k = 1, \cdots, K-1$, are the individuals' characteristics, γ_k are the corresponding parameters, which measure the impact of each of the factors on the change in utility (Bateman et al., 2002), and γ is a constant term.

After estimating the parameters, the mean WTP can be calculated by

$$mean\, WTP = \int_0^\infty [1 - F(t)]\, dt = \int_0^\infty \pi^y(t)dt, \tag{2.16}$$

where $F(t)$ is the cumulative distribution function (cdf) of the WTP. In the case of the probit model, the mean WTP can be calculated as

$$mean\, WTP = \int_0^\infty \Phi(\hat{\alpha} - \hat{\beta}t)dt, \tag{2.17}$$

where $\hat{\alpha}$ and $\hat{\beta}$ are the estimated parameters. The untruncated mean WTP values for each model are given in Table 2.1. Note that with the log-logistic model, the untruncated mean diverges to infinity if the parameter estimate of β is equal to or lower than one.

If the model includes individual characteristics as in Eq. (2.15), $\hat{\alpha}$ in Eq. (2.17) is normally a "grand constant," which is calculated as the sum of the estimated constant plus the product of the other independent variables and their respective means (Loomis et al., 2000).

Eq. (2.16) indicates that there is some portion of the respondents who have a WTP greater than their income. This violates standard economic theory, because no one can pay more than his/her income. To avoid this, the truncated mean can be used as an alternative. In the early years of SBDC-CV research, this was calculated as

$$\int_0^{t_{max}} [1 - F(t)]\, dt. \tag{2.18}$$

The highest bid in the survey (denoted as t_{max} in the equation above) is used as the upper limit of integration in most cases.

However, as Boyle et al. (1988) suggested,

$$F(t_{max}) < \lim_{x \to \infty} F(x) = 1, \tag{2.19}$$

and thus, Eq. (2.18) is not a correct indication of the expected WTP. Alternatively, Boyle et al. (1988) proposed a normalization procedure under the assumption that

$$F(t) = 0, \quad \text{if} \quad t > t_{max}, \tag{2.20}$$

the proper truncated mean WTP is calculated as

$$\int_0^{t_{max}} \left[\frac{1 - F(t)}{F(t_{max})}\right] dt. \tag{2.21}$$

The median WTP, which is said to be generally a more robust measure of central tendency (Hanemann, 1984), can be calculated as

$$median\, WTP = F^{-1}(0.5). \tag{2.22}$$

The median WTP values for each model discussed above are also given in Table 2.1.

Table 2.1 *Acceptance probability, mean, and median*

Distribution	$P^y(t)$	Mean	Median
Normal	$\Phi(\alpha - \beta t)$	$\frac{\alpha}{\beta}$	$\frac{\alpha}{\beta}$
Logistic	$\frac{1}{1+\exp(-\alpha+\beta t)}$	$\frac{\alpha}{\beta}$	$\frac{\alpha}{\beta}$
Log-normal	$\Phi(\alpha - \beta \ln(t))$	$\exp(\frac{\alpha}{\beta})\exp(\frac{1}{2\beta^2})$	$\exp(\frac{\alpha}{\beta})$
Log-logistic	$\frac{1}{1+\exp(-\alpha+\beta\ln(t))}$	$\exp(\frac{\alpha}{\beta})\Gamma(1-\frac{1}{\beta})\Gamma(1+\frac{1}{\beta})$	$\exp(\frac{\alpha}{\beta})$
Weibull	$\exp\{-\exp(-\alpha+\beta\ln(t))\}$	$\exp(\frac{\alpha}{\beta})\Gamma(1-\frac{1}{\beta})$	$\frac{\exp(\frac{\alpha}{\beta})}{(\ln(2))^{\beta}}$

Source: Carson and Hanemann (2005) and Hanemann and Kanninen (1999).

Note: In the log-logistic and Weibull distributions, finite median exists when $\beta > 1$. Otherwise it becomes infinity.

Parametric estimation of DBDC data

Data from the DBDC format can be analyzed by an expanded version of the approach presented above for SBDC (Hanemann et al., 1991).

Let the first bid be t_n and the second bid t_n^U if respondent n answers "yes" to the first question, and t_n^L otherwise. Further, let respondent n's maximum WTP be denoted by y_n^*.

The probability that respondent n answers "yes" to the first and second questions is given by

$$\begin{aligned} P^{yy}(t_n, t_n^U) &= \Pr(t_n \leq y_n^*, t_n^U \leq y_n^*) \\ &= \Pr(t_n^U \leq y_n^*) \\ &= 1 - F(t_n^U). \end{aligned} \tag{2.23}$$

Similarly, the probability that respondent n answers "no" to the first and second questions is equal to

$$\begin{aligned} P^{nn}(t_n, t_n^L) &= \Pr(y_n^* \leq t_n, y_n^* \leq t_n^L) \\ &= \Pr(y_n^* \leq t_n^L) \\ &= F(t_n^L). \end{aligned} \tag{2.24}$$

The probability that respondent n answers "yes" to the first question and "no" to the second, or "no" to the first question and "yes" to the second is given, respectively, by

$$\begin{aligned} P^{yn}(t_n, t_n^U) &= \Pr(t_n \leq y_n^*, y_n^* < t_n^U) \\ &= \Pr(t_n \leq y_n^* < t_n^U) \\ &= F(t_n^U) - F(t_n) \end{aligned} \tag{2.25}$$

and

$$\begin{aligned} P^{ny}(t_n, t_n^L) &= \Pr(y_n^* < t_n, y_n^* \geq t_n^L) \\ &= \Pr(t_n^L \leq y_n^* < t_n) \\ &= F(t_n) - F(t_n^L). \end{aligned} \tag{2.26}$$

Therefore, for a given sample of N independent observations, the log-likelihood function can be written as

$$\begin{aligned} \ln L = \sum_{n=1}^{N} \big[& d_n^{yy} \ln\{P^{yy}(t_n, t_n^U)\} + d_n^{nn} \ln\{P^{nn}(t_n, t_n^L)\} \\ & + d_n^{yn} \ln\{P^{yn}(t_n, t_n^U)\} + d_n^{ny} \ln\{P^{ny}(t_n, t_n^L)\}\big], \end{aligned} \tag{2.27}$$

where d_n^{yy}, d_n^{nn}, d_n^{yn}, and d_n^{ny} are binary-valued indicator variables. For example, d_n^{yy} is equal to one if respondent n answers "yes" to both the first bid t_n and the second bid t_n^U, and zero otherwise.

With some assumption on $F(\ )$, parameters can be estimated using the maximum likelihood techniques as is the case with SBDC-CV data.

For example, if $F()$ is the standard normal cdf, Eqs. (2.23)–(2.26) become

$$\begin{aligned} P^{yy}(t_n, t_n^U) &= 1 - F(t_n^U) \\ &= 1 - \Phi(-\alpha + \beta t_n^U) \\ &= \Phi(\alpha - \beta t_n^U) \end{aligned} \tag{2.28}$$

$$\begin{aligned} P^{nn}(t_n, t_n^L) &= F(t_n^L) \\ &= \Phi(-\alpha + \beta t_n^L) \\ &= 1 - \Phi(\alpha - \beta t_n^L) \end{aligned} \tag{2.29}$$

$$\begin{aligned} P^{yn}(t_n, t_n^U) &= F(t_n^U) - F(t_n) \\ &= \Phi(-\alpha + \beta t_n^U) - \Phi(-\alpha + \beta t_n) \\ &= \Phi(\alpha - \beta t_n) - \Phi(\alpha - \beta t_n^U) \end{aligned} \tag{2.30}$$

$$\begin{aligned} P^{ny}(t_n, t_n^L) &= F(t_n) - F(t_n^L) \\ &= \Phi(-\alpha + \beta t_n) - \Phi(-\alpha + \beta t_n^L) \\ &= \Phi(\alpha - \beta t_n^L) - \Phi(\alpha - \beta t_n). \end{aligned} \tag{2.31}$$

So, the corresponding log-likelihood function can be written as

$$\begin{aligned}\ln L = \sum_{n=1}^{N} \big[& d_n^{yy} \ln\{\Phi(\alpha - \beta t_n^U)\} + d_n^{nn} \ln\{1 - \Phi(\alpha - \beta t_n^L)\} \\ & + d_n^{yn} \ln\{\Phi(\alpha - \beta t_n) - \Phi(\alpha - \beta t_n^U)\} \\ & + d_n^{ny} \ln\{\Phi(\alpha - \beta t_n^L) - \Phi(\alpha - \beta t_n)\}\big]. \end{aligned} \tag{2.32}$$

In the case where $F()$ is the standard logistic cdf, Eqs. (2.23)–(2.26) become

$$\begin{aligned} P^{yy}(t_n, t_n^U) &= 1 - F(t_n^U) \\ &= \frac{1}{\exp(-\alpha + \beta t_n^U)} \\ &= \exp(\alpha - \beta t_n^U) \end{aligned} \tag{2.33}$$

$$\begin{aligned} P^{nn}(t_n, t_n^L) &= F(t_n^L) \\ &= 1 - \frac{1}{\exp(-\alpha + \beta t_n^L)} \\ &= 1 - \exp(\alpha - \beta t_n^L) \end{aligned} \tag{2.34}$$

$$\begin{aligned} P^{yn}(t_n, t_n^U) &= F(t_n^U) - F(t_n) \\ &= \left\{1 - \frac{1}{\exp(-\alpha + \beta t_n^U)}\right\} - \left\{1 - \frac{1}{\exp(-\alpha + \beta t_n)}\right\} \\ &= \exp(\alpha - \beta t_n) - \exp(\alpha - \beta t_n^U) \end{aligned} \tag{2.35}$$

$$\begin{aligned} P^{ny}(t_n, t_n^L) &= F(t_n) - F(t_n^L) \\ &= \left\{1 - \frac{1}{\exp(-\alpha + \beta t_n)}\right\} - \left\{1 - \frac{1}{\exp(-\alpha + \beta t_n^L)}\right\} \\ &= \exp(\alpha - \beta t_n^L) - \exp(\alpha - \beta t_n). \end{aligned} \tag{2.36}$$

The corresponding log-likelihood function is given by

$$\begin{aligned}\ln L = \sum_{n=1}^{N} \big[& d_n^{yy} \ln\left\{\exp(\alpha - \beta t_n^U)\right\} + d_n^{nn} \ln\left\{1 - \exp(\alpha - \beta t_n^L)\right\} \\ & + d_n^{yn} \ln\left\{\exp(\alpha - \beta t_n) - \exp(\alpha - \beta t_n^U)\right\} \\ & + d_n^{ny} \ln\left\{\exp(\alpha - \beta t_n^L) - \exp(\alpha - \beta t_n)\right\}\big]. \end{aligned} \tag{2.37}$$

2.2.4 Nonparametric estimation method

As seen above, parametric estimation requires some assumption on the distribution of WTP, which suggests that the chosen distribution may be incorrect. On the other hand, the nonparametric approach requires no such assumption.

In general, nonparametric techniques provide a purely empirical approach to estimating the survival function, which gives the probability of observing a particular value of WTP or more (Bateman et al., 2002).

Nonparametric estimation of SBDC data

Nonparametric estimation of SBDC binary data, proposed by Kriström (1990), is explained below.

Suppose there are J levels of bids and each bid level is denoted as $t_j, j = 0, \cdots, J$, where t_0 is assumed to be zero.

First, we calculate the proportion of "yes" responses for each bid t_j as

$$\widehat{S}(t_j) = \frac{n_j}{N_j}, \tag{2.38}$$

where N_j is the number of respondents in the sub-sample responding to bid level t_j, and n_j is the number of respondents answering "yes" to the CV question. $\widehat{S}(t_0)$ is assumed to be one, meaning that the WTP is non-negative.

If $\widehat{S}(t_j)$ forms a monotonic non-increasing sequence, this sequence provides a distribution free maximum likelihood estimator of the probability for acceptance (Kriström, 1990).

If the sequence is non-monotonic, the next step is to apply the pooled adjacent violators algorithm (PAVA). If $\widehat{S}(t_j) < \widehat{S}(t_{j+1})$ for some j, then $\widehat{S}(t_j)$ and $\widehat{S}(t_{j+1})$ are replaced by

$$\frac{n_j + n_{j+1}}{N_j + N_{j+1}}. \tag{2.39}$$

The procedure is repeated until the sequence is monotonic.

Now that we have a sequence of estimated probabilities, the next challenge is to obtain a survival function from these probabilities. Kriström (1990) used linear interpolation to obtain a survival function and calculated the mean WTP by the area under the dashed line in Figure 2.1. This estimator corresponds to what is known as the Spearman–Kärber estimator and can be calculated as (Carson and Hanemann, 2005)

$$mean\, WTP_{SK} = \frac{1}{2}\sum_{j=1}^{J+1} \left(\widehat{S}(t_j) + \widehat{S}(t_{j-1})\right)(t_j - t_{j-1}). \tag{2.40}$$

However, the $meanWTP_{SK}$ in Eq. (2.40) cannot be calculated without the upper distribution endpoint t_{J+1}, such that $\widehat{S}(t_{J+1}) = 0$ (Boman et al., 1999). Normally, however, the endpoint cannot be determined from the data. Although Kriström (1990) and Boman and Bostedt (1999) used an endpoint found by extrapolation, there is no general consensus on how to obtain the endpoint. Boman et al. (1999) examined the sensitivity of estimation to the choice of the endpoint.

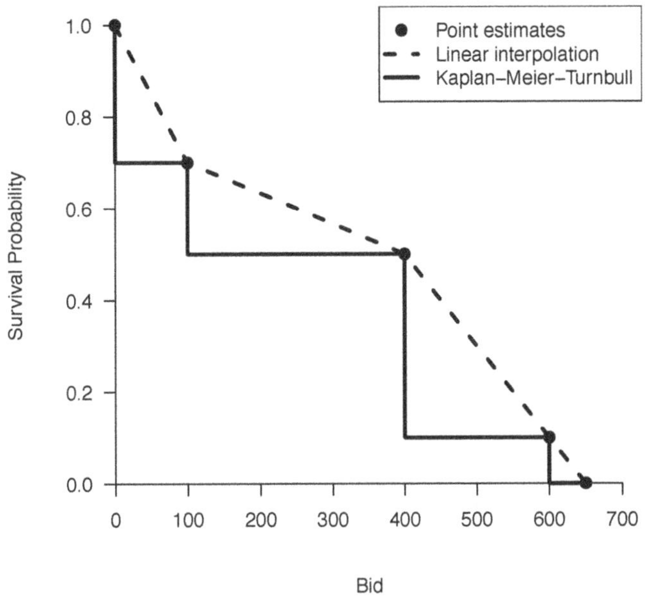

Figure 2.1 *Survival functions obtained by nonparametric methods.*

There are several other ways of interpolating between each probability. Of these, the most "conservative" approach is frequently used. In this approach, the survival function between two successive bid levels, t_{j-1} and t_j, is interpolated with $\widehat{S}(t_j)$, the lower of the two probabilities (the solid lines in Figure 2.1). The mean WTP, corresponding to the area under the step function, is calculated as

$$mean\, WTP_{KMT} = \sum_{j=1}^{J} \widehat{S}(t_j)(t_j - t_{j-1}). \tag{2.41}$$

This is known as the Kaplan–Meier–Turnbull estimator (Carson and Hanemann, 2005). The median WTP can be obtained in each approach. In Kriström's approach, the median WTP corresponds to the point at which the value of the survival function is 0.5. The survival function of the Kaplan–Meier–Turnbull method is a step function, so it cannot yield a point estimate of the median. Instead, it gives the interval in which the median WTP falls (Carson et al., 1992).

The Kaplan–Meier–Turnbull estimates of mean and median can be taken as the lower bounds for the statistics. Since they give the minimum value of the mean or median that is consistent with the sample data, this estimation is an indispensable step in the analysis of CV data (Bateman et al., 2002).

Nonparametric estimation of DBDC data

The basic concept of nonparametric estimation of DBDC data is mostly the same as that of the single-bounded case. The difference lies in how to calculate the proportion of "yes" responses for each bid t_j.

Since the data from the double-bounded format are interval data, there can be an "overlap" between the intervals corresponding to the respondents' answers. Carson and Steinberg (1990) applied a modification of the Kaplan–Meier estimator proposed by Turnbull (1976) to double-bounded dataAfter obtaining a sequence of survival probabilities, the mean and median WTP values can be calculated in a similar manner to the single-bounded case.

2.3 An R package for analyzing SBDC and DBDC CV data

*2.3.1 Overview of package **DCchoice***

To analyze SBDC and DBDC CV data, we prepared an add-on package called **DCchoice**. The package comprises two parts, one for parametric estimation and the other for nonparametric estimation.

The parametric estimation part of the package contains two main functions, `sbchoice()` and `dbchoice()`. As the names of the functions suggest, `sbchoice()` is designed for modeling SBDC data, and `dbchoice()` for DBDC data. The estimation procedure in `sbchoice()` uses the `glm()` function in the **stats** package, whereas `dbchoice()` numerically optimizes the hard-coded log-likelihood function. Both functions estimate parametric models (currently, logistic, normal, log-logistic, and log-normal distributions) of the respondents' choice behavior. In addition, **DCchoice** contains two functions, `krCI()` and `bootCI()`, for constructing the confidence intervals of the WTP estimates in `sbchoice()` and `dbchoice()`. The `krCI()` function formulates the confidence intervals based on the simulation method proposed by Krinsky and Robb (1986, 1990), whereas `bootCI()` uses the bootstrap method to construct them.

For nonparametric estimation of WTP, **DCchoice** is equipped with three main functions, `kristrom()`, `turnbull.sb()`, and `turnbull.db()`. Whereas `kristrom()` estimates survival probabilities of the bids using the method introduced by Kriström (1990), `turnbull.sb()` and `turnbull.db()` return the Kaplan–Meier–Turnbull estimates of survival probabilities for SBDC and DBDC data, respectively. In estimating the Kaplan–Meier–Turnbull survival probabilities, `turnbull.sb()` and `turnbull.db()` rely on the `icfit()` function, which is part of the **interval** package (Fay and Shaw, 2010).

All of these functions are examined in detail in Sections 2.4 and 2.5 for parametric and nonparametric models, respectively.

2.3.2 *Installing the **DCchoice** package*

The package **DCchoice** is currently available from R-Forge, a platform for the development of R packages and R-related projects. The information such as a development status and the source code of **DCchoice** is available at `https://R-Forge.R-project.org/projects/dcchoice/`.

To install **DCchoice** and other dependent packages simultaneously, execute the following lines of code in the R console:

```
R> install.packages("DCchoice",
     repos = c("http://R-Forge.R-project.org",
          "http://www.bioconductor.org/packages/release/bioc",
          "http://cran.rstudio.com/"),
     dep = TRUE)
```

For Mac and Unix/Linux users, add the option argument `type="source"` to the `install.packages()` function. Since **DCchoice** is available only from R-Forge, and other dependent packages are spreading over a couple of repositories, it is important to set the `repos` option correctly. Otherwise, the automated installation of dependent packages may fail.

Package installation in R is usually carried out by the Package Installer accessible from the Packages & Data menu of the R GUI (for Windows and Mac OS). However, the availability of `DCchoice` and package dependency prevent the user from using the Package Installer. Therefore, please follow the instruction above for installation of **DCchoice**.

2.3.3 *Loading the package*

To be able to use the functionality in **DCchoice**, the user needs to load the package in advance. This is done in the R console by executing the command:

```
R> library(DCchoice)   # loading the package
Loading required package: MASS
Loading required package: Ecdat
Loading required package: interval
Loading required package: survival
Loading required package: splines
Loading required package: perm
Loading required package: Icens
Loading required package: MLEcens
```

This command additionally loads other dependent packages necessary for **DCchoice** to work correctly. Once **DCchoice** and other dependent packages have been successfully loaded and several lines have been output giving the names of the dependent packages loaded, the R prompt is once again displayed on the console.

2.3.4 Preparing datasets

In the remainder of this chapter, we use several datasets from large-scale CV surveys. Some of these are bundled in **DCchoice**. We also use a dataset contained in the **Ecdat** package (Croissant, 2014), which is a collection of data related to various econometric studies. Before going into the detail of these datasets, some remarks are given regarding how the results of a DBDC survey are stored in a machine readable format.

A DBDC survey asks a respondent twice whether he/she accepts the suggested value. If the respondent does not wish to accept it in the first stage, the next bid is decreased by a certain amount and presented to the same person to see whether it is now acceptable. In contrast, if the respondent answers "yes" to the first bid, the second one is increased and again presented to the respondent.

A respondent's stated preference to the bids, therefore, assuming that the "don't know" answer is not allowed, falls into one of four categories, that is,

- "yes to the first bid, and yes to the second one" (yy),
- "yes to the first bid, and no to the second one" (yn),
- "no to the first bid, and yes to the second one" (ny), or
- "no to the first bid, and no to the second one" (nn).

There are at least two ways of recording respondents' preferences for analysis in R.

The first approach, which is based on the structure of a DBDC questionnaire, uses two binary variables and two bid variables. Each binary variable, say `R1` or `R2`, records `1` for "yes" and `0` for "no." Each bid variable, `bid1` or `bid2`, stores the actual amount of the bid presented to the respondent at each stage. This is how the Albemarle–Pamlico sounds survey data, which are bundled as `AP` in **DCchoice**, are recorded. Details of the data are illustrated in Section 2.3.5.

The second method uses a single categorical variable (say, `answers`) with four levels to denote the preference status. The variable keeps the respondent's stated preference "as is," that is, one of {`yy`, `yn`, `ny`, `nn`} or simply {`1`, `2`, `3`, and `4`}. The `NaturalPark` data in the **Ecdat** package, illustrated in Section 2.3.5, use this convention. An aggregated form of this method is used in the double-bounded version of the *Exxon Valdez* Oil Spill data, which are bundled as `CarsonDB` and examined in Section 2.3.5. In addition, these two datasets keep the increased and decreased bids (e.g., `bidh` and `bidl`) for the second stage instead of using a single variable (`bid2`) for the actual second bid. To use the datasets in **DCchoice**, these variables are required to be transformed in the way suggested by the first method. The R code for such a transformation is given in Section 2.3.5 together with the explanation of the datasets.

For single-bounded survey data, it is sufficient to organize a data frame object by the first approach with `R1` and `bid1` as well as other covariates.

2.3.5 *Example datasets*

*Datasets included in **DCchoice***

DCchoice contains example datasets that have been analyzed in existing articles. In what follows, brief explanations of the included data are given.

The Albemarle–Pamlico sounds survey data

The original data[3] were based on a telephone survey regarding quality improvements in the Albemarle–Pamlico sounds, North Carolina. The data have been intensively analyzed, for instance, by Whitehead (1995) and Whitehead et al. (1998) in different contexts.

The original data have 1077 observations and include the bids and responses of the DBDC survey. Various socio-demographic characteristics were also collected in the survey. After removing observations with missing entries, a subset ($N = 721$) of the original data is bundled with **DCchoice** as a data frame object called `AP`.

`AP` consists of the responses to the CV questions as well as a small number of socio-demographic characteristics. The first three observations of `AP` are as follows:

```
R> data(AP)        # loading the Albemarle-Pamlico sounds CV data
R> head(AP, 3)   # showing the first three rows
  bid1 bid2 R1 R2 income work age female married
1  300  600  1  1   5000    0  24      0       0
2  300  150  0  0  40000    1  30      1       1
3  400  200  0  0   5000    0  70      0       0
```

The first two columns correspond to the first and second bids presented to the respondents (labeled `bid1` and `bid2`, respectively) while the next two correspond to the responses to the bids (`R1` and `R2`, respectively). Five socio-demographic variables are included, namely, `income`; denoting annual household income in 1995 dollars; `work`, which is a dummy variable equal to 1 if the respondent is a full-time worker and 0 otherwise; `age` denoting the age of the respondent; `female`, which is a dummy variable equal to 1 if the respondent is female and 0 otherwise; and `married`, representing a dummy variable equal to 1 if the respondent is married and 0 otherwise.

The Exxon Valdez Oil Spill data

In 1989, the tanker *Exxon Valdez* ran aground on rocks in a bay in Alaska, U.S. The oil compartment of the tanker was damaged in the grounding, causing a

[3]The complete dataset is downloadable at
`http://www.appstate.edu/\~{}whiteheadjc/research/data/ap/`.

massive oil spill in the area. The oil spill was recognized as "one of the major environmental disasters in U.S. history" (Carson et al., 2003).

A large-scale contingent valuation study was conducted to assess the damage caused by the *Exxon Valdez* Oil Spill. Details of the study were reported in Carson et al. (1992). Among them, Table A-15 in Appendix C.1 summarizes responses to the first stage questions in the DBDC survey, while Tables A-16 and A-17 contain information on the second stage questions.

These tables are included in **DCchoice** as two data frame objects: `CarsonSB`, corresponding to Table A-15 in Carson et al. (1992), and `CarsonDB`, which contains the information given in Tables A-16 and A-17 as well as A-15.

```
R> data(CarsonSB)
R> CarsonSB
   T1   Y   N
1  10 178  86
2  30 138 129
3  60 129 126
4 120  88 169
```

```
R> data(CarsonDB)
R> CarsonDB
   T1  TU TL  yy yn ny  nn
1  10  30  5 119 59  8  78
2  30  60 10  69 69 31  98
3  60 120 30  54 75 25 101
4 120 250 60  35 53 30 139
```

As the names suggest, `CarsonSB` is designed for use with SBDC examples, while `CarsonDB` is for double-bounded ones.

To use the datasets with functions in `DCchoice`, the contingency tables must be converted into data frame objects with individual observations. We explain how to do this by creating a data frame object, `sb.data`, from `CarsonSB`.

The process requires the following rather tricky programs in R.

```
R> n <- rowSums(CarsonSB[, -1])
R> sb.data <- data.frame(
     bid = c(rep(CarsonSB$T1[1], n[1]),
             rep(CarsonSB$T1[2], n[2]),
             rep(CarsonSB$T1[3], n[3]),
             rep(CarsonSB$T1[4], n[4])),
     R1 = c(rep(1, CarsonSB$Y[1]), rep(0, CarsonSB$N[1]),
            rep(1, CarsonSB$Y[2]), rep(0, CarsonSB$N[2]),
            rep(1, CarsonSB$Y[3]), rep(0, CarsonSB$N[3]),
            rep(1, CarsonSB$Y[4]), rep(0, CarsonSB$N[4]))
   )
```

The first line of the code above counts the numbers of respondents for each bid using the `rowSums()` function, and saves the result in a new object `n`. In the remaining lines of code, a new data frame called `sb.data` is created using `data.frame()` with two components, `bid` and `R1`. The code fragment beginning with `bid =` creates component `bid` of which the first `n[1]` entries are equal to `CarsonSB$T1[1]`, the subsequent `n[2]` entries are equal to `CarsonSB$T1[2]`, and so on, where the number in square brackets (`[]`) denotes the position of the entry and `CarsonSB$T1` is a vector of bids. The next part of the code, starting with `R1 =`, records the response to the bid, namely, `1` for "`yes`" and `0` otherwise, for each entry in `bid`. The length of `bid` and `R1` is equal to the number of respondents.

The created object `sb.data` has two columns, namely, the bid and the response, as shown below.

```
R> head(sb.data)    # showing the first six rows
  bid R1
1  10  1
2  10  1
3  10  1
4  10  1
5  10  1
6  10  1
```

The contingency table created from `sb.data` is identical to object `CarsonSB`, except for the order of the columns and the column names, that is,

```
R> table(sb.data)
     R1
bid     0   1
  10   86 178
  30  129 138
  60  126 129
  120 169  88
```

Transformation of object `CarsonDB` can be carried out similarly; the code is summarized in the appendix to this chapter.

Kriström's data

The data frame object `KR` consists of the responses to the SBDC survey of 900 Swedish citizens regarding preservation of the eleven virgin forests in Sweden. The respondents were divided into sub-groups, and each sub-group was asked to accept or reject a different level of bids for the cost of the preservation. The response rate was 67%, with the sample size $N = 562$. See Kriström (1990) for more details.

Object `KR` has the following form, which is similar to that of `sb.data`:

```
R> data(KR)
R> head(KR, 3)
```

```
  bid1 R1
1  100  1
2  100  1
3  100  1
```

`KR` is used in Section 2.5.1 to replicate the results in Kriström (1990).

The `NaturalPark` *data*

The **Ecdat** package (Croissant, 2014), on which **DCchoice** depends, is a collection of datasets related to econometric research. Among the data therein, `NaturalPark` consists of WTP and other socio-demographic variables of 312 individuals regarding the preservation of a natural park in Portugal.[4]

Let us take a look at how the observations are organized. `NaturalPark` is loaded[5] by executing `data()` with the name of the **Ecdat** package.

```
R> data(NaturalPark, package = "Ecdat")
R> head(NaturalPark, 3)
  bid1 bidh bidl answers age    sex income
1    6   18    3      yy   1 female      2
2   48  120   24      yn   2   male      1
3   48  120   24      yn   2 female      3
```

The first three rows show that `NaturalPark` consists of seven variables, of which the first three correspond to bids and the fourth records the answers to both the first and second bids in "`yy/nn`" format. Socio-demographic characteristics are included in the last three columns.

To use `NaturalPark` data in **DCchoice**, two different data transformations are needed. One is used to create the variable containing the second bid (`bid2`), which is actually presented to the respondents, while the other rearranges `answers` into `R1` and `R2`, that is, the answers to the first and second stage bids, respectively. In R, it is straightforward to create `bid2` from `bidl`, `bidh` and `answers` in `NaturalPark`. This can be done, for example, using the `ifelse()` function.

```
R> NP <- NaturalPark          # a new object for transformation
R> NP$bid2 <- ifelse(NP$answers == "yy" | NP$answers == "yn",
       NP$bidh, NP$bidl)
```

A call to the logical function `ifelse()` takes the form `ifelse(condition, true, false)`, and returns `true` if `condition` is true, and `false` otherwise. The vertical bar `|` is an "or" operator in a logical expression. The prefix `NP$` is a command that makes it possible to access each variable contained in `NP` directly.

[4]The survey was conducted in 1997 by Paulo Nunes for his Ph.D. thesis, and a subset of the results was supplied for illustrative use in Verbeek (2004).

[5]To load the `NaturalPark` data, **Ecdat** must be installed in advance. Since the **DCchoice** package depends on **Ecdat**, the latter is installed at the same time as **DCchoice**.

Now the new variable bid2 is concatenated in the last column of NP, whose entries are organized in such a way that the corresponding item is taken from bidh if the response to the first bid is yes (that is, NP$answers == "yy" or "yn"), and from bidl otherwise.

In R it is also not difficult to create the two binary variables, R1 and R2, out of the categorical variable answers in NP. Using ifelse() again, the following two lines of code perform this task.

```
R> NP$R1 <- ifelse(NP$answers == "yy" | NP$answers == "yn",
                   1, 0)
R> NP$R2 <- ifelse(NP$answers == "yy" | NP$answers == "ny",
                   1, 0)
```

Finally, for use in the subsequent sections, bid1 and bid2, which are the bids at the first and second stages, respectively, are further transformed into natural logs and are saved, respectively, as Lbid1 and Lbid2 in NP.

```
R> NP$Lbid1 <- log(NP$bid1)
R> NP$Lbid2 <- log(NP$bid2)
```

The contents of NP are shown below.

```
R> head(NP, 3)
  bid1 bidh bidl answers age    sex income bid2 R1 R2 Lbid1
1    6   18    3      yy   1 female      2   18  1  1 1.792
2   48  120   24      yn   2   male      1  120  1  0 3.871
3   48  120   24      yn   2 female      3  120  1  0 3.871
  Lbid2
1 2.890
2 4.787
3 4.787
```

The columns to the right of income have been appended to NP. The new variables that were created or transformed from existing ones are clearly shown in the above.

2.4 Parametric estimation of WTP

2.4.1 Estimating WTP with SBDC data

The function for analyzing SBDC data is sbchoice(). A generic call to sbchoice() is given by

```
sbchoice(formula,
         data,
         dist = "log-logistic")
```

Argument formula defines a relation between the response variable and covariates, while data is mandatory since it specifies the data frame object containing the variables in the model. The third argument, dist, sets the error distribution in the model.

Assuming that the response variable is `R1`, the covariates are `var1` and `var2`, the bid variable is `bid1`, and all the variables are stored in a single data frame object, a typical `formula` is defined as

```
R1 ~ var1 + var2 | bid1
```

It is important to note that the bid variable `bid1` is placed after the vertical bar (`|`). The formula for a model with only an intercept and the bid variable is defined as

```
R1 ~ 1 | bid1
```

Inclusion of the bid variable (`bid1` in the current example) is mandatory in the formula for `sbchoice()`. This model corresponds to the null or restricted model in calculating the pseudo R^2.

If one is keen to estimate a model with intercept only, `glm()` can be used with the formula given by `R1 ~ 1`. Consult the help on `glm()` in the **stats** package for the details.

The error distribution defaults to `dist="log-logistic"`. This suggests that the bid variable in `formula` must be expressed as a natural log. The same is true if `dist="log-normal"`or `dist="weibull"`. However, a bid without any transformation (other than a change in unit) must be specified if `dist="logistic"` or `dist="normal"` is used. This is because the difference between normal and log-normal models or between logistic and log-logistic ones lies in how the bid variable is incorporated into the model. For the Weibull model, the log of the bid variable is customarily used.

The output from `sbchoice()` is an S3 class object "`sbchoice`". The extractor function `summary()` is available for summarizing the fitted model objects of the "`sbchoice`" class. The function `summary()` returns the table of estimates and the associated inferential statistics such as their standard errors and the z-values. In addition, `summary()` computes four different WTP estimates: the expected WTP, the mean WTP truncated at the maximum bid with and without adjustment, and the median WTP.

Examples

We examine how `sbchoice()` and `summary()` work by estimating log-logistic and log-normal models for the Natural Park data.

Example code: A log-logistic model of the Natural Park data

Suppose that a log-logistic model is to be estimated using the `NP` data, which were transformed from `NaturalPark` in Section 2.3.5. The response variable is called `R1`, and the names of the covariates are `sex`, `age`, and `income`. The log of the bid is saved as `Lbid1`. The model is estimated and the results are saved in the "`sbchoice`" class object `sb.logit` by the command

```
R> sb.logit <- sbchoice(R1 ~ sex + age + income | Lbid1,
                        data = NP)
```

The contents of `sb.logit` are summarized by the extractor function `summary()` as follows.

```
R> summary(sb.logit)
Call:
sbchoice(formula = R1 ~ sex + age + income | Lbid1,
    data = NP)

Formula:
R1 ~ sex + age + income | Lbid1

Coefficients:
             Estimate Std. Error z value Pr(>|z|)
(Intercept)  2.35184    0.64755   3.632 0.000281
sexfemale   -0.61025    0.25045  -2.437 0.014827
age         -0.37090    0.08546  -4.340 1.42e-05
income       0.26038    0.10566   2.464 0.013730
log(bid)    -0.46244    0.16212  -2.852 0.004338

Distribution: log-logistic
Number of Obs.: 312
log-likelihood: -190
pseudo-R^2: 0.1142 , adjusted pseudo-R^2: 0.0909
LR statistic: 49.1 on 4 DF, p-value: 0.000
AIC: 390.578 , BIC: 409.293

Iterations: 4
Convergence: TRUE

WTP estimates:
 Mean : Inf (because of |beta_bid| < 1)
 Mean : 26.3 (truncated at the maximum bid)
 Mean : 46.9 (truncated at the maximum bid with adjustment)
 Median : 28.1
```

In the first few lines output by `summary(sb.logit)`, the call to the function and the formula are shown. These are followed by the table of coefficients with relevant inferential statistics. It should be noted that the estimate on the bid variable is always reported at the bottom of the table. The name label of the bid variable is fixed: `log(bid)` if the assumed distribution is either `dist = "log-logistic"` (default) or `dist = "log-normal"`, and `BID` if `dist = "logistic"` or `dist = "normal"`. Recall that `sex` is a two-level factor variable as displayed in Section 2.3.5. Factor variables are internally processed as dummy variables. In the current example, `sex = male` is defined as the base level.

The output from `summary(sb.logit)` also gives the assumed error distribution, the number of observations, and Akaike's and the Bayesian information criterion (labeled as `AIC` and `BIC`, respectively), as well as the value of the log-likelihood at the estimates, pseudo-R^2, and the likelihood ratio (LR) statistic for the current model. The two lines labeled `Iterations` and `Convergence` report the number of iterations required to achieve convergence, and the convergence status, respectively, of the numerical optimization. Finally, the mean and median estimates of WTP are presented.

In the summarized output, four different WTP estimates are calculated. The first is the expected WTP based on the unmodified error distribution, while the second is the truncated mean WTP, computed under the assumption that the error distribution is truncated at the maximum bid. The third estimate is the truncated mean WTP adjusted by the method of Boyle et al. (1988). The final estimate is the median WTP. Analytic solutions for the mean and median WTPs are summarized in Table 2.1. The expected WTP in the log-logistic model becomes infinite if the absolute value of the parameter estimate on the bid is smaller than or equal to 1. All the mean WTP estimates are computed by numerical integration according to Eqs. (2.16), (2.18), and (2.21) for the expected WTP, the truncated mean, and the adjusted mean, respectively.

In the example code above, the estimate of the mean WTP becomes infinite (`Mean : Inf`) because the necessary condition for a finite mean WTP, $|\beta_{bid}| > 1$, is violated; that is, we have $\hat{\beta}_{Lbid} = -0.462$. The truncated mean is usually finite, and is estimated to be 26.3 euros. The median WTP is 28.1 euros. It is left for the user to decide whether the mean, the truncated mean with/without adjustment, or the median WTP is used when the representative WTP for the non-marketed good in question is reported.

Example code: A log-normal model of the Natural Park data

The log-normal model is estimated by explicitly stating `dist = "log-normal"` in `sbchoice()` and the outcome thereof summarized by

```
R> sb.normal <- sbchoice(R1 ~ sex + age + income | Lbid1,
                         data = NP,
                         dist = "log-normal")
R> summary(sb.normal)
Call:
sbchoice(formula = R1 ~ sex + age + income | Lbid1,
    data = NP, dist = "log-normal")

Formula:
R1 ~ sex + age + income | Lbid1

Coefficients:
            Estimate Std. Error z value Pr(>|z|)
```

```
(Intercept)  1.43387    0.38527   3.722 0.000198
sexfemale   -0.36925    0.15125  -2.441 0.014633
age         -0.22497    0.05097  -4.414 1.02e-05
income       0.15219    0.06213   2.450 0.014302
log(bid)    -0.28076    0.09794  -2.867 0.004149

Distribution: log-normal
Number of Obs.: 312
log-likelihood: -190
pseudo-R^2: 0.1134 , adjusted pseudo-R^2: 0.0901
LR statistic: 48.7 on 4 DF, p-value: 0.000
AIC: 390.932 , BIC: 409.647

Iterations: 4
Convergence: TRUE

WTP estimates:
 Mean : 15573
 Mean : 26.2 (truncated at the maximum bid)
 Mean : 46.5 (truncated at the maximum bid with adjustment)
 Median : 27.4
```

The output of the log-normal model has the same structure as that of the log-logistic one. The discrepancy between the estimated truncated mean and median WTPs of the log-normal model and the log-logistic model is moderate. In contrast to the log-logistic model, the expected WTP in the normal or log-normal model remains finite regardless of the value of the estimate of the bid variable. However, the estimate of the expected WTP can be unrealistically large, as is shown in the output.

The `summary()` function literally summarizes the contents of `sb.logit` or `sb.normal`. The components contained in the latter object can be seen by executing

```
R> names(sb.normal)
 [1] "coefficients" "call"         "formula"
 [4] "glm.out"      "glm.null"     "distribution"
 [7] "covariates"   "bid"          "nobs"
[10] "yn"           "data.name"
```

As displayed, `sb.normal` is made up of 11 components. The function `summary()` arranges these into an output table. Individual items are accessible by using the $ operator. For example,

```
R> sb.normal$coefficients
(Intercept)   sexfemale         age      income    log(bid)
      1.434      -0.369      -0.225       0.152      -0.281
```

returns the coefficient estimates.[6] Furthermore, the object created by `summary()` itself is a list object with the following components:

```
R> names(summary(sb.normal))
 [1] "coefficients"      "call"
 [3] "formula"           "glm.out"
 [5] "glm.null"          "distribution"
 [7] "covariates"        "bid"
 [9] "nobs"              "yn"
[11] "data.name"         "glm.summary"
[13] "glm.null.summary"  "loglik"
[15] "loglik.null"       "medianWTP"
[17] "meanWTP"           "trunc.meanWTP"
[19] "adj.trunc.meanWTP" "psdR2"
[21] "adjpsdR2"          "LR.test"
[23] "AIC"
```

The items listed above can, therefore, be extracted using the prefix `summary(sb.normal)$`. Extracting the information criterion, for example, can be achieved by

```
R> summary(sb.normal)$AIC
   AIC    BIC
390.93 409.65
```

Extracting the coefficient table is slightly more tricky but can be done by

```
R> summary(sb.normal)$glm.summary$coef
            Estimate Std. Error z value   Pr(>|z|)
(Intercept)  1.43387   0.385267  3.7217 1.9785e-04
sexfemale   -0.36925   0.151250 -2.4413 1.4633e-02
age         -0.22497   0.050973 -4.4135 1.0171e-05
income       0.15219   0.062128  2.4496 1.4302e-02
log(bid)    -0.28076   0.097942 -2.8666 4.1487e-03
```

This trickiness comes from the fact that `summary(sb.normal)$glm.summary` itself is a list object with nested components among which the coefficient table is saved.

2.4.2 Estimating WTP with DBDC data

Next we turn our attention to carrying out a study on DBDC data. The function for DBDC data is called `dbchoice()`. Usage of this function is almost identical to that of `sbchoice()`. A generic call to `dbchoice()` is given by

[6]The generic function `coefficients()` can also be used to extract the estimates. That is, `coefficients(sb.normal)` returns the vector of estimates. However, this does not work for the Weibull model because the estimates are saved in a different way.

Full details of the components can be found by executing in the R console "`R> ?sbchoice`" for the `sbchoice()` function and "`R> ?summary.sbchoice`" for the `summary` method for `sbchoice` class objects.

```
dbchoice(formula,
         data,
         dist = "log-logistic",
         par = NULL)
```

The call to `dbchoice()` differs from that to `sbchoice()` in the syntax for `formula`, and in the additional argument, `par`, which gives the vector of starting values for numerical optimization.

In DBDC data, four possible outcomes are required to describe the responses if a "don't know" answer is not allowed. One way of recording this situation is to define bid and response variables for each stage. The `formula` for `dbchoice()` therefore, has to include these four variables as well as the covariates. A typical `formula` has the form

```
R1 + R2 ~ var1 + var2 | bid1 + bid2
```

where `R1` and `R2` are the responses to the first and second bids, denoted by `bid1` and `bid2`, respectively, and `var1` and `var2` are the covariates.

The additional argument, `par`, determines the starting values for the parameters in the maximization of the log-likelihood function. It defaults to `NULL`, which means that the initial parameter vector is internally found using a probit model employing only `bid1` as the bid variable. The log-likelihood function for the DBDC data is given by Eq. (2.27), which is a composition of the four distinct choice statuses.

There is no guarantee that the likelihood function is unimodal, meaning that the optimization may fail with a particular choice of starting values. The argument `par` gives the user the opportunity for setting alternative starting values, which may result in successful convergence.

Practical usage of the functions is illustrated by estimating log-logistic and log-normal models for the double-bounded version of the Albemarle–Pamlico sounds survey data.

Example code: Log-logistic and log-normal models for the Albemarle–Pamlico sounds data

The usage of `dbchoice()` is exemplified by fitting log-logistic and log-normal models to the double-bounded version of the Albemarle–Pamlico sounds data in `AP`. In Section 2.3.5, we saw that `AP` included five variables other than the bid and response variables. We assume that the choice of the suggested WTP is a function of `female`, `age`, `married`, and the log of `income`.

Since `dbchoice()` defaults to the log-logistic distribution, executing

```
R> data(AP)
R> db.logit <- dbchoice(R1 + R2 ~ female + age + married +
                         log(income) | log(bid1) + log(bid2),
                         data = AP)
```

estimates a log-logistic model. The log-normal model is estimated by explicitly adding the argument `dist = "log-normal"`.

```
R> db.normal <- dbchoice(R1 + R2 ~ female + age + married +
                         log(income) | log(bid1) + log(bid2),
                         data = AP,
                         dist = "log-normal",
                         par = db.logit$coefficients)
```

In the log-normal model, the estimates of the parameters in the logistic model are supplied as an initial parameter vector. This is because the optimization failed when using the internally determined initial parameter vector. To see the unfavorable outcome, the user can execute the command given above, without the parameter `par = db.logit$coefficients` in the log-normal model.

A normal or logistic model can be estimated with `dist = "normal"` or `dist = "normal"`, respectively. But then, `log(bid1)` and `log(bid2)` in the model formula must be replaced by `bid1` and `bid2`, respectively.

In addition, the Weibull model can be estimated with `dist = "weibull"`. It is warned that it only accepts the log of the bid variables.

The function `dbchoice()` returns an S3 class object "`dbchoice`". Relevant information can be extracted by `summary()`. The output of `summary()` for the "`dbchoice`" class object is the same as that for "`sbchoice`".

The components saved in the "`dbchoice`" class objects, `db.logit` for the log-logistic model and `db.normal` for the log-normal model, are summarized by `summary(db.logistic)` and `summary(db.normal)`, respectively. Summarized outcomes are available through the extractor function `summary()`:

```
R> summary(db.logit)
Call:
dbchoice(formula = R1 + R2 ~ female + age + married +
    log(income) | log(bid1) + log(bid2), data = AP)

Formula:
R1 + R2 ~ female + age + married + log(income) | log(bid1) +
    log(bid2)

Coefficients:
              Estimate Std. Error z value  Pr(>|z|)
(Intercept)  3.866525   1.133178   3.412  0.000645
female       0.346016   0.149000   2.322  0.020219
age         -0.024648   0.004743  -5.197 < 2.2e-16
married     -0.356667   0.165999  -2.149  0.031666
log.income.  0.298830   0.108978   2.742  0.006105
log(bid)    -1.271842   0.064474 -19.727 < 2.2e-16

Distribution: log-logistic
Number of Obs.: 721
Log-likelihood: -883.7191
```

```
LR statistic: 53.173 on 4 DF, p-value: 0.000
AIC: 1779.4381 , BIC: 1806.9219

Iterations: 44 13
Convergence: TRUE

WTP estimates:
 Mean : 371.53
 Mean : 181.57 (truncated at the maximum bid)
 Mean : 193.42 (truncated at the maximum bid with adjustment)
 Median: 93.577
```

for the log-logistic model and

```
R> summary(db.normal)
Call:
dbchoice(formula = R1 + R2 ~ female + age + married +
    log(income) | log(bid1) + log(bid2), data = AP,
    dist = "log-normal", par = db.logit$coefficients)

Formula:
R1 + R2 ~ female + age + married + log(income) | log(bid1) +
    log(bid2)

Coefficients:
              Estimate Std. Error z value  Pr(>|z|)
(Intercept)  2.304996   0.661228   3.486  0.000490
female       0.190012   0.087879   2.162  0.030602
age         -0.013967   0.002714  -5.146 < 2.2e-16
married     -0.225372   0.098014  -2.299  0.021483
log.income.  0.175737   0.063670   2.760  0.005778
log(bid)    -0.759193   0.036193 -20.976 < 2.2e-16

Distribution: log-normal
Number of Obs.: 721
Log-likelihood: -879.4200

LR statistic: 49.693 on 4 DF, p-value: 0.000
AIC: 1770.8400 , BIC: 1798.3238

Iterations: 75 11
Convergence: TRUE

WTP estimates:
 Mean : 219.28
 Mean : 177.8 (truncated at the maximum bid)
```

```
Mean : 187.23 (truncated at the maximum bid with adjustment)
Median: 92.1
```

for the log-normal model.

The output above shows the relevant information of the estimated model, such as the value of the log-likelihood at the estimates, the table of estimates, and the usual inferential statistics. Finally, the mean, the truncated mean with/without adjustment, and the median estimates of the WTP are reported at the end of the output.

Before interpreting the outcome, the user should heed the convergence status of the optimization reported under the label `Convergence`. The indicator `Convergence: TRUE` means that the optimization converged successfully. If for some reason the optimization failed, `Convergence: FALSE` is returned and the message `"The optimization did not converge"` is shown before the coefficient table. Recall that the log-likelihood function is given by Eq. (2.27), which is highly nonlinear in its parameters. It is therefore, more likely to encounter an unsuccessful convergence.

The `Convergence` information is preceded by the line with the label `Iterations` followed by two integers. The first entry is the number of calls (iterations) to the objective function and the second is the number of evaluations of the gradient of the objective function.

The output of `dbchoice()` is a list with various components, which can be obtained using the command

```
R> names(db.logit)
 [1] "f.stage"      "dbchoice"     "coefficients"
 [4] "call"         "formula"      "Hessian"
 [7] "distribution" "loglik"       "convergence"
[10] "niter"        "nobs"         "covariates"
[13] "bid"          "yn"           "data.name"
```

This implies that the user may retrieve particular outcomes by, say,

```
R> coefficients(db.logit)
(Intercept)      female        age    married log.income.
   3.866525    0.346016  -0.024648  -0.356667    0.298830
   log(bid)
  -1.271842
R> db.logit$loglik
[1] -883.72
```

and the like. All of the components can be extracted with the prefix `db.logit$`. Moreover, `summary(db.logit)` returns another list with various components:

```
R> sum_db.logit <- summary(db.logit)
R> names(sum_db.logit)
```

```
 [1] "f.stage"             "dbchoice"
 [3] "coefficients"        "call"
 [5] "formula"             "Hessian"
 [7] "distribution"        "loglik"
 [9] "convergence"         "niter"
[11] "nobs"                "covariates"
[13] "bid"                 "yn"
[15] "data.name"           "medianWTP"
[17] "meanWTP"             "trunc.meanWTP"
[19] "adj.trunc.meanWTP"   "LR.test"
[21] "coef"                "AIC"
```

The content of most of the components is self-evident according to their names.[7] One could, for instance, display the coefficient table by

```
R> sum_db.logit$coef
             Estimate Std. Error z value Pr(>|z|)
(Intercept)    3.8665    1.13318    3.41 0.000645
female         0.3460    0.14900    2.32 0.020219
age           -0.0246    0.00474   -5.20 0.000000
married       -0.3567    0.16600   -2.15 0.031666
log.income.    0.2988    0.10898    2.74 0.006105
log(bid)      -1.2718    0.06447  -19.73 0.000000
```

as well as export it to an external file by

```
R> write.table(sum_db.logit$coef, file = "dbout.txt",
       quote = FALSE)
```

The table has been saved in `"dbout.txt"`, which is created in the current working directory. The output directory may be independently specified by setting `file = "c:/foo/dbout.txt"`, for example.

To estimate a model without covariates other than a constant and the (log of the) bid, the following can be executed:

```
R> db.logit.null <- dbchoice(R1 + R2 ~ 1 | log(bid1) +
                             log(bid2), data = AP)
```

A constant term is denoted by `1`. This model is often called the restricted or null model, meaning that none of the excluded covariates affects the choice behavior of WTP. The null model commonly serves as a benchmark to diagnose the unrestricted model. The `dbchoice()` function internally estimates the null model to obtain and the likelihood ratio (LR) test statistic for the unrestricted model. The pseudo-R^2 measure is not computed because $\ln(0)$ appears in the log-likelihood function of the intercept-only models (see Eqs. (2.32) or (2.37) with $\alpha = \beta = 0$).

[7] Again, full details of the components can be found by executing "`R> ?summary.dbchoice`" in the R console.

2.4.3 Constructing confidence intervals

In summarizing the output of an estimated model, it is desirable to construct confidence intervals for the various WTP estimates. In Chapter 1 we briefly reviewed the theoretical background of three such major procedures, namely, the delta and the bootstrap methods, and the method by Krinsky and Robb (1986, 1990). While users need to write their own code to perform the delta method at the expense of programming time, the **DCchoice** package is equipped with two functions for the latter two methods. Both these methods rely on simulation techniques with different settings.

The Krinsky and Robb method

Function `krCI()` uses the Krinsky and Robb simulation method (Krinsky and Robb, 1986, 1990) to construct confidence intervals for the four different WTPs, estimated by functions `sbchoice()` or `dbchoice()`. A call to the function is given as

```
krCI(obj, nsim = 1000, CI = 0.95)
```

where `obj` is an object of either the "`sbchoice`" or "`dbchoice`" class, `nsim` is the number of draws of the parameters, and `CI` is the percentile of the confidence intervals to be estimated. The user can change the values of `nsim` and `CI`.

The `krCI()` function returns an S3 class list object "`krCI`", which includes a table of the simulated confidence intervals as well as vectors containing the simulated WTPs.

The bootstrap method

Function `bootCI()` carries out the bootstrapping and returns the bootstrap confidence intervals. The call to `bootCI()` is analogous to that of `krCI()` and is given by

```
bootCI(obj, nboot = 1000, CI = 0.95)
```

where `obj` is an object of either the "`sbchoice`" or "`dbchoice`" class, `nboot` is the number of bootstrap iterations, and `CI` is the percentile of the confidence intervals to be constructed. The function defaults to returning 95% confidence intervals with 1000 bootstrap replications. Although the user can change the confidence level and the number of bootstrap iterations, the default settings are adequate.[8]

[8] It should be mentioned that raising the number of bootstrap iterations can result in a huge increase in computation time with little improvement in the confidence interval estimate. The computation time is an increasing (exponential) function of the sample size and the number of covariates. For the Natural Park example, which is a simple model with a minimum number of covariates and the bid, and with $N = 312$, it took about 10 seconds using the default settings to return the outcome on a 32-bit PC with a 3.33-GHz Intel Core i7 processor. However, it took more than two minutes to carry out the same task with `nboot = 10000`, which yielded no marked change in confidence interval estimates.

Example code

The confidence intervals for the WTP estimates are obtained using `krCI()` for the Krinsky and Robb method, and `bootCI()` for the bootstrap method. For the Natural Park example, these are obtained as follows:

```
R> sblogit.krCI <- krCI(sb.logit)
R> sblogit.krCI
the Krinsky and Robb simulated confidence intervals
                        Estimate       LB       UB
Mean                         Inf -999.000 -999.000
truncated Mean            26.327   23.277   29.033
adjusted truncated Mean   46.892   35.687   61.325
Median                    28.138   15.604  105.437
```

and

```
R> sblogit.boCI <- bootCI(sb.logit)
R> sblogit.boCI
the bootstrap confidence intervals
                        Estimate       LB       UB
Mean                         Inf -999.000 -999.000
truncated Mean            26.327   23.536   29.367
adjusted truncated Mean   46.892   36.233   62.194
Median                    28.138   16.548   92.329
```

Each function returns a table containing the four different estimates of WTP, and the lower and upper bounds (denoted by `LB` and `UB`, respectively) of the specified (the default is 95%) confidence intervals. The WTP estimates are those reported by `summary(sb.logit)`. If the parameter estimate of the bid does not satisfy the condition for the existence of a finite mean WTP, both functions return the interval $[-999, -999]$. Indeed in the Natural Park example, $|\hat{\beta}_{bid}| = |-0.462| < 1$, and therefore, the upper and lower bounds are coerced to `-999`.

To be precise, `krCI()` and `bootCI()` return S3 class objects "`krCI`" and "`bootCI`", respectively. The components contained in these class objects, `sblogit.krCI` and `sblogit.bootCI`, are given below.

```
R> names(sblogit.krCI)
[1] "mWTP"       "tr.mWTP"      "adj.tr.mWTP"
[4] "medWTP"     "out"
R> names(sblogit.boCI)
[1] "mWTP"       "tr.mWTP"      "adj.tr.mWTP"
[4] "medWTP"     "out"
```

As can be seen above, `sblogit.krCI` and `sblogit.bootCI` are both list objects containing components with the same names. Simulated outcomes are saved in the vector components: `mWTP` for the mean WTP, `tr.mWTP` for the mean WTP truncated at the maximum bid, `adj.tr.mWTP` for the truncated mean with adjustment, and `medWTP` for the median WTP. The table of simulated confidence intervals is actually stored in `out`. It should be mentioned that if $\hat{\beta}_{bid}$ does not satisfy the finite mean WTP condition, the simulated outcomes for the median WTP become unstable in the sense that outliers frequently appear in both tails in addition to the mean WTP being infinite and thus without confidence bounds.

Using these vector components, characteristics of the empirical distributions of the simulated WTPs can be investigated. For instance, the bootstrapped mean and standard errors of the truncated mean WTP can be obtained by

```
R> mean(sblogit.boCI$tr.mWTP)
[1] 26.3
R> sd(sblogit.boCI$tr.mWTP)
[1] 1.57
```

The same operations can be carried out for `sblogit.krCI`.

Moreover, the user can, for example, draw a histogram of the empirical distribution of the simulated WTP. In the case of the truncated mean WTP in `sblogit.krCI`, the following command can be used to create the empirical distribution, which is illustrated in Figure 2.2.

```
R> hist(sblogit.krCI$tr.mWTP, main="",
           xlab="truncated mean WTP")
```

2.5 Nonparametric estimation of WTP

For nonparametric estimation of WTP, **DCchoice** offers three functions, namely, `kristrom()` for Kriström's method, and `turnbull.sb()` and `turnbull.db()` for the Kaplan–Meier–Turnbull method with SBDC and DBDC data, respectively.

In what follows, the usage of each function is demonstrated by estimating empirical survival probabilities from accompanying datasets. The next two sections, Subsections 2.5.1 and 2.5.2, deal with SBDC data, while Subsection 2.5.3 focuses on double-bounded data. At the end of Subsection 2.5.1, we attempt to replicate the results of Kriström (1990).

2.5.1 Kriström's nonparametric estimation of SBDC data

Nonparametric estimation of WTP was first conducted by Kriström (1990). For the method proposed by Kriström (1990), **DCchoice** includes a function called `kristrom()`. The generic call to `kristrom()` is simple; assuming that the response and bid are stored in `R1` and `bid1`, respectively, the call is given by

```
    kristrom(R1 ~ bid1, data)
```

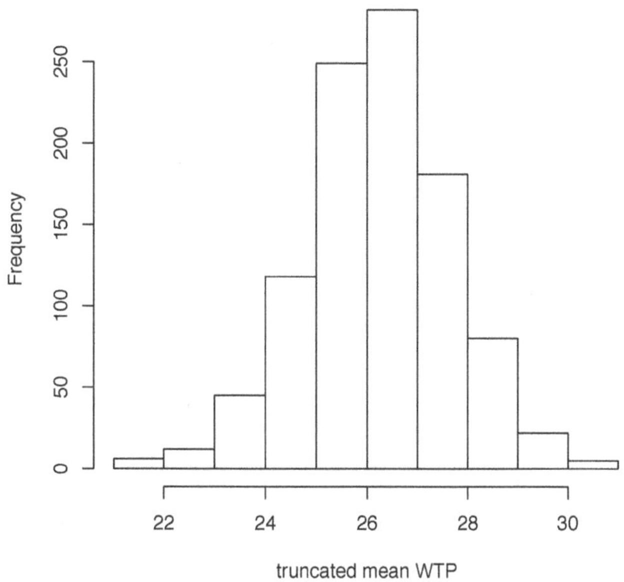

Figure 2.2 *Empirical distribution of the simulated truncated mean WTP.*

Since the method is designed for nonparametric estimation of WTP, there is no argument for the distribution. Moreover, no covariates are included in the formula. The `kristrom()` function returns an S3 class object "`kristrom`". The result of `kristrom()` is summarized by the generic extractor function `summary()`. An empirical survival function can be visualized by the generic function `plot()`.

Basic procedures for `kristrom()` can be exemplified by applying it to the Albemarle–Pamlico sounds data saved in `AP`. Assuming that `AP` has already been loaded by the command `load(AP)`, the following two lines of code estimate and summarize Kriström's survival probabilities for the Albemarle–Pamlico sounds data.

```
R> kr.AP <- kristrom(R1 ~ bid1, data = AP)
R> summary(kr.AP)
Survival probability:
   Upper   Prob.
1      0  1.0000
2    100  0.4011
3    200  0.4011
4    300  0.3256
```

```
5     400  0.2294
6     Inf  0.0000

WTP estimates:
 Mean: 135.71040  (Kaplan-Meier)
 Mean: 201.60279  (Spearman-Karber)
 Median: 83.48018
```

The summarized output contains a table with the estimates of the survival probabilities and three different WTP estimates. Intervals are defined as $(B_l, B_u]$ in the estimation where B_l and B_u denote the lower and upper bounds, respectively. To save space, only the values of B_u are displayed in the summarized output. The estimate of a survival probability in each row indicates $\mathsf{Pr}(B_l < WTP \leq B_u)$. In addition, the first and last rows (corresponding to the intervals $[0, 0]$ and $(400, \infty)$, and whose probabilities are always one and zero, respectively) are attached for convenience.

The three reported WTP estimates are the Kaplan–Meier mean estimate, the Spearman–Kärber mean estimate, and the median estimate. In principle, these estimates are based on the area under the empirical survival function. The `kristrom()` function computes the rectangular area under the empirical survival function up to the maximum bid for the Kaplan–Meier estimates, whereas for the Spearman–Kärber estimates it computes the area under the survival function up to the x-intercept.[9] Section 2.2.4 provides details of these computations.

It is useful to show the estimated survival probabilities visually. The empirical survival function of the Albemarle–Pamlico sounds data can be graphically illustrated by

```
R> plot(kr.AP)
```

The graph is shown in Figure 2.3. The appearance of the graph, such as the title and line thickness, can be changed by including the graphic parameters listed in Table 2.2 as arguments for `plot()`. If further modification of the plot is necessary, the estimates and corresponding upper and lower bounds can be retrieved by `kr.AP$estimates`, and then various graphic functions provided in R can be used to display these.

Replicating Kriström's results

We attempted to replicate the results for the nonparametric estimation of WTP as well as the parametric estimates (the logit-model) as given in Kriström (1990). The data used by Kriström (1990) are included in **DCchoice** as a data frame object called `KR`. As discussed in Section 2.3.5, `KR` consists of

[9]We use linear extrapolation between the two points on the bid-probability plane for the last interval. If these two points give the same survival probability, the slope of the line passing through the points for the second last interval is used to find the x-intercept. The value of the x-intercept is retrieved by `summary(kr.AP)$x.icpt`.

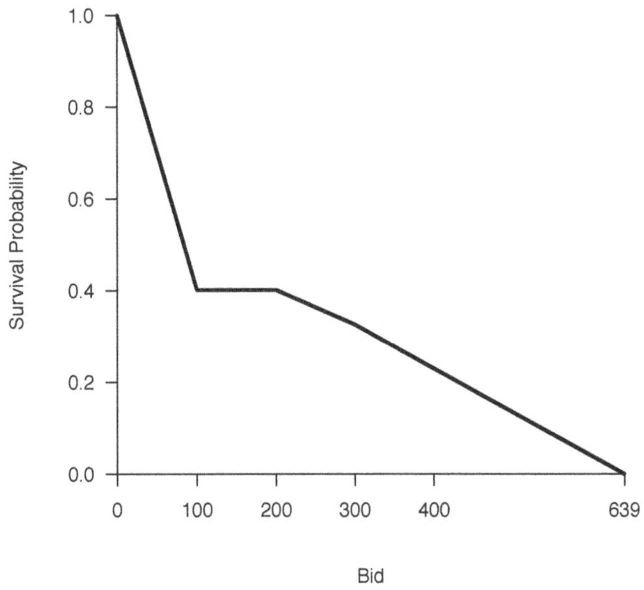

Figure 2.3 *Empirical survival function for the Albemarle–Pamlico sounds data (single-bounded) based on Kriström's nonparametric method.*

two variables, `bid1` and `R1`, where `bid1` is the bid presented to the respondents and `R1` is the response to the bid.

First we create a frequency table of the data and resulting probabilities, thereby replicating Table 1 in Kriström (1990).

```
R> data(KR)
R> tab <- table(KR)
R> prop.tab <- prop.table(tab, margin = 1)
R> tab <- addmargins(tab, margin = 2)
R> tab <- cbind(tab[, -1], round(prop.tab[, 2], 3))
R> colnames(tab) <- c("yes", "total", "yes/total")
```

The first line loads the data `KR`, which is then transformed into a contingency table by the second line. The third line creates the proportion table by row. Row-wise totals are added to the last column of the contingency table `tab` in the fourth line. The fifth line attaches the second column of the proportion table (corresponding to the ratio of the number of "`yes`" responses to the total number of respondents) to the contingency table. Note that the first column of `tab` is excluded in the fourth line. The last line adds column names to `tab`.

The replicated table is shown below:

Table 2.2 *Graphic parameters for* `plot()`

Argument	Description and default value
`main`	Main title of the plot. By default, no main title is displayed.
`sub`	Sub-title of the plot. By default, no sub-title is displayed.
`xlab`	x-axis label of the plot. By default, `xlab = "Bid"` is used. Setting `xlab = ""` displays no x-axis label.
`ylab`	y-axis label of the plot. By default, `ylab = "Survival Probability"` is used. Setting `ylab = ""` displays no y-axis label.
`lwd`	Line width for the plot. By default, `lwd = 3` is used.
`lty`	Line type for the plot. By default, `lty = 1` is used.

```
R> tab
     yes total yes/total
100   51    60     0.850
400   29    52     0.558
700   33    60     0.550
1000  31    57     0.544
1500  25    64     0.391
2000  16    56     0.286
2500  21    53     0.396
3000  16    53     0.302
5000  21    62     0.339
7000   5    45     0.111
```

This is a partial replication of Table 1 in Kriström (1990). The first column of the table contains the amount of the bid suggested for each subsample. The second and third columns include the number of respondents responding "yes" to the bid and the sizes of the subsamples, respectively. The difference between `tab` and Table 1 in Kriström (1990) appears in the last column as the fraction of yes answers to the suggested bid (that is, `yes/total`). The fraction is not monotonically decreasing as the amount of the bid increases. This was the main motivation for Kriström (1990).

The following two lines of code are used to estimate Kriström's survival probabilities and summarize the result.

```
R> kr.example <- kristrom(R1 ~ bid1, data = KR)
R> summary(kr.example)
Survival probability:
    Upper   Prob.
1       0  1.0000
2     100  0.8500
3     400  0.5577
```

```
4      700  0.5500
5     1000  0.5439
6     1500  0.3906
7     2000  0.3394
8     2500  0.3394
9     3000  0.3217
10    5000  0.3217
11    7000  0.1111
12     Inf  0.0000

WTP estimates:
 Mean: 2141.79768  (Kaplan-Meier)
 Mean: 2519.99054  (Spearman-Karber)
 Median: 1143.11270
```

It can be seen that the estimates of acceptance probabilities in the third column of the output table above (in the column labeled `Prob.`) coincide with those reported in the third column ($p(\text{“}yes\text{”})$) of Table 1 in Kriström (1990). The Spearman–Kärber estimate in the current example is 2520, which is roughly the same as that reported in Table 2 in Kriström (1990). The discrepancy lies in the way the x-intercept is found.

The empirical survival function of Kriström's data can be depicted using the generic function `plot(kr.example)` to create Figure 2.4. Note that Figure 2.4 is not exactly the same as that shown in Kriström (1990).

Kriström (1990) also estimated a log-logistic model for the purpose of comparison. The result is easily replicated by `sbchoice()` with only an intercept as the covariate.

```
R> kr.sb <- sbchoice(R1 ~ 1 | log(bid1), data = KR)
R> summary(kr.sb)
Call:
sbchoice(formula = R1 ~ 1 | log(bid1), data = KR)

Formula:
R1 ~ 1 | log(bid1)

Coefficients:
            Estimate Std. Error z value Pr(>|z|)
(Intercept)  4.63716    0.61741   7.511 5.88e-14
log(bid)    -0.68003    0.08493  -8.007 1.17e-15

Distribution: log-logistic
Number of Obs.: 562
log-likelihood: -346.936
pseudo-R^2: 0.1004 , adjusted pseudo-R^2: 0.0952
LR statistic: 77.456 on 1 DF, p-value: 0.000
```

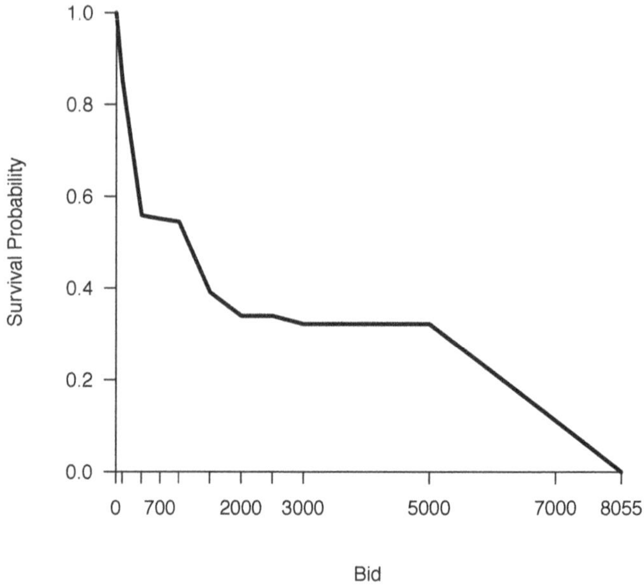

Figure 2.4 *Kriström's nonparametric estimation of the empirical survival function.*

```
AIC: 697.87247 , BIC: 706.53547

Iterations: 4
Convergence: TRUE

WTP estimates:
 Mean : Inf (because of |beta_bid| < 1)
 Mean : 2367.43 (truncated at the maximum bid)
 Mean : 2960.86 (truncated at the maximum bid with adjustment)
 Median : 915.076
```

The above code reproduces the results presented in Table 2 in Kriström (1990). As Kriström (1990) pointed out, $\hat{\beta}_{Lbid} = -0.68$ does not satisfy the sufficient condition $|\beta_{Lbid}| > 1$ for a finite estimate of mean WTP. The truncated mean WTP estimate without adjustment is 2367 SEK, which is reasonably close to 2372 SEK in Kriström (1990). The remaining small deviation could be attributed to differences in precision of the parameter estimates or in the algorithms used in the numerical integral.

2.5.2 Kaplan–Meier–Turnbull estimation of SBDC data

Another technique for estimating survival probabilities for SBDC data is the Kaplan–Meier–Turnbull method. This can be implemented using function `turnbull.sb()` in **DCchoice**, which returns an S3 class object "`turnbull`".

The generic call to `turnbull.sb()` is identical to that of `kristrom()`, that is,

```
turnbull.sb(formula, data)
```

Argument `formula` is defined as `R1 ~ bid1`, in the same way as for `kristrom()`.

To illustrate the usage of `turnbull.sb()`, two examples are given. The first, once again, uses the Natural Park data, while the second uses the *Exxon Valdez* data from Carson et al. (1992).

Example code using the Natural Park data

The Kaplan–Meier–Turnbull survival probability estimates of the Natural Park data are estimated and summarized by

```
R> tb.NP <- turnbull.sb(R1 ~ bid1, data = NP)
R> summary(tb.NP)
Survival probability:
   Upper   Prob.
1      0  1.0000
2      6  0.6579
3     12  0.5584
4     24  0.5122
5     48  0.4675
6    Inf  0.0000

WTP estimates:
 Mean: 24.66514  (Kaplan-Meier)
 Mean: 26.80324  (Spearman-Karber)
 Median in: [     24 ,      48 ]
```

The appearance of the summarized output is similar to that of the "`kristrom`" object. However, there are two differences. The first is that the median estimate is reported as an interval owing to the fact that it is not possible to report a point estimate of the median because the Kaplan–Meier–Turnbull survival function is a step function, which means that there are infinitely many candidate points for the median estimate. The second difference lies in how the Spearman–Kärber mean estimate is computed. It is the area under Kriström's type of empirical survival function. However, in the Kaplan–Meier–Turnbull case, this empirical survival function is truncated at the maximum bid, and so is the area. Therefore, while the empirical survival probabilities and the Kaplan–Meier–Turnbull mean estimate are approximately the same as the ones in the output from the `kristrom()` function, the Spearman–Kärber mean estimate is smaller by the triangular area beyond the maximum bid.[10]

The generic function `plot()` can be used to draw the estimated survival function as follows:

```
R> plot(tb.NP)
```

with the outcome shown in Figure 2.5. The arguments listed in Table 2.2 are also applicable here for changing the main title, labels of axes, and so on.

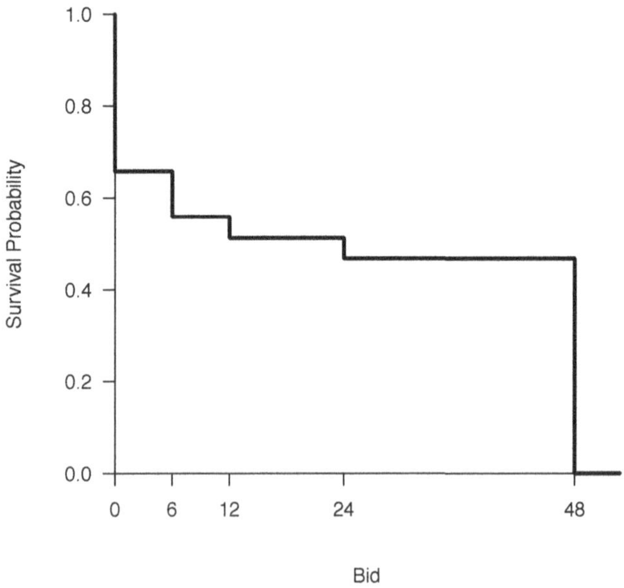

Figure 2.5 *Kaplan–Meier–Turnbull estimate of the empirical survival function for the Natural Park data.*

Example code using the Exxon Valdez Oil Spill data

The accompanying data `CarsonSB` is a reproduction of Table A-15 in Appendix C.1 in Carson et al. (1992). We transformed `CarsonSB` into `sb.data` in

[10]As discussed in Section 2.2.4, there seems to be no unified approach for determining the endpoint of bids at which the acceptance probability is zero. Therefore, we leave the decision of how the area is dealt with, to the user. In practice, once the endpoint, bid_{end}, is found, it is straightforward to compute the triangular area by $0.5(bid_{end} - bid_{max})P_{max}$, where bid_{max} is the maximum bid and P_{max} is the acceptance probability for bid_{max}, both of which are reported in the summarized output. In the Natural Park example, $bid_{max} = 48$ and $P_{max} = 0.468$.

Section 2.3.5 for use with the functions in **DCchoice**. We now apply `turnbull.db` to `sb.data`.

A call to the function is executed by the first line of code given below, while the summarized output is generated by the second line.

```
R> carson.sb <- turnbull.sb(R1 ~ bid, data = sb.data)
R> summary(carson.sb)
Survival probability:
   Upper   Prob.
1      0  1.0000
2     10  0.6742
3     30  0.5169
4     60  0.5059
5    120  0.3424
6    Inf  0.0000

WTP estimates:
 Mean: 52.80072  (Kaplan-Meier)
 Mean: 61.07206  (Spearman-Karber)
 Median in: [      60 ,      120 ]
```

We see that the Kaplan–Meier and Spearman–Kärber estimates of the mean WTP are USD 52.8 and 61.1, respectively. The median WTP is estimated to lie in the interval between USD 60 and 120 . The estimated survival function is displayed in Figure 2.6.

2.5.3 Kaplan–Meier–Turnbull estimation of DBDC data

For the Kaplan–Meier–Turnbull estimation of DBDC data, package **DCchoice** includes the function `turnbull.db()`, which returns an S3 class object "`turnbull`". The call to `turnbull.db()` is the same as that for `turnbull.sb()`, that is,

```
turnbull.db(formula, data)
```

However, the `formula` argument must be defined in a different way. As mentioned previously, DBDC behavior is fully described by two binary variables for the responses and bids, respectively. The `formula` argument for `turnbull.db()` is therefore, expressed as

```
R1 + R2 ~ bid1 + bid2
```

where `R1` and `R2` on the left-hand side of the tilde operator (~) denote the binary variables for the responses, and `bid1` and `bid2` on the right-hand side denote the first and second bids, respectively.

Example code using the Natural Park data

Once again, we make use of the Natural Park data. In Section 2.3.5 we created the necessary variables to use the functions in **DCchoice**, and saved them in `NP`. The names of the variables in `NP` are

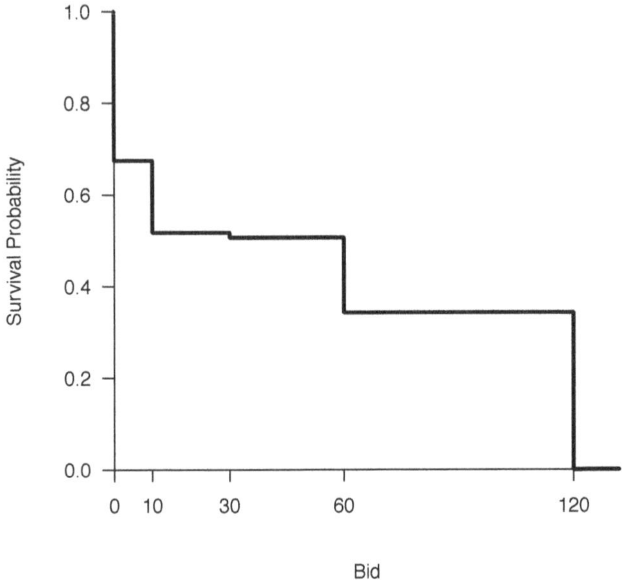

Figure 2.6 *Kaplan–Meier–Turnbull estimate of the empirical survival function for the Exxon Valdez data.*

```
R> names(NP)
 [1] "bid1"    "bidh"    "bidl"    "answers" "age"
 [6] "sex"     "income"  "bid2"    "R1"      "R2"
[11] "Lbid1"   "Lbid2"
```

Of the variables in NP, four are used: R1 and R2 for the responses, and bid1 and bid2, for the bids. A call to turnbull.db() is given by

```
R> trdb.NP <- turnbull.db(R1 + R2 ~ bid1 + bid2, data = NP)
R> summary(trdb.NP)
Survival probability:
  Upper    Prob.
1     0  1.00000
2     3  0.69831
3     6  0.65896
4    12  0.59574
5    18  0.41667
6    24  0.41667
7    48  0.22549
8   120  0.01879
```

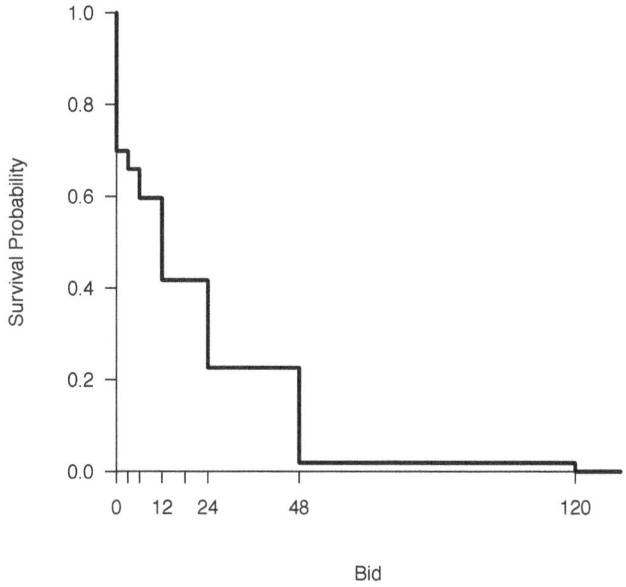

Figure 2.7 *Kaplan–Meier–Turnbull estimate of the empirical survival function for the Natural Park data (double-bounded).*

```
9     Inf  0.00000

WTP estimates:
 Mean: 19.41095  (Kaplan-Meier)
 Mean: 30.38469  (Spearman-Karber)
 Median in: [      12 ,      18 ]
```

The estimated survival function is plotted in Figure 2.7.

Example code using the double-bounded version of the Exxon Valdez Oil Spill data

We continue working on the *Exxon Valdez Oil Spill* data, but this time using the double-bounded version, which is contained in the data frame object `CarsonDB` in **DCchoice**. `CarsonDB` is in the form of a contingency table and must be transformed before using the data in `turnbull.db()`. The conversion procedure is described in the appendix to this chapter, with the converted data saved in the data frame object `db.data`. The first three observations in `db.data` are

```
R> head(db.data, 3)
  bid1 bid2 R1 R2 bid2U bid2L
1   10   30  1  1    30     5
2   10   30  1  1    30     5
3   10   30  1  1    30     5
```

The columns labeled `bid1` and `bid2` include the first-and second-stage bids shown to the respondents, respectively, while `R1` and `R2` are vectors of the responses to `bid1` and `bid2`, respectively. For completeness, `bid2U` and `bid2L` are also included, but these are not used in the analysis.

The Kaplan–Meier–Turnbull estimates of the survival probabilities are obtained and summarized by

```
R> carson.db <- turnbull.db(R1 + R2 ~ bid1 + bid2,
                            data = db.data)
R> summary(carson.db)
Survival probability:
  Upper    Prob.
1     0  1.00000
2     5  0.72057
3    10  0.69191
4    30  0.54122
5    60  0.38390
6   120  0.22070
7   250  0.08778
8   Inf  0.00000

WTP estimates:
 Mean: 54.05765  (Kaplan-Meier)
 Mean: 72.23059  (Spearman-Karber)
 Median in: [     30 ,      60 ]
```

The estimated survival probabilities are slightly different from the results reported in Carson et al. (1992, 2003) owing to the fact that Carson et al. (1992, 2003) used the data adjusted for possible selection bias in response to the survey, whereas we have used unadjusted data.[11] The empirical survival function is illustrated using `plot(carson.db)` in Figure 2.8.

2.6 Concluding remarks

In this chapter, we have dealt with the SBDC- and DBDC-CV methods. After reviewing the literature in various fields of application, analytical frameworks were discussed for both parametric and nonparametric setups. The R package **DCchoice** for analyzing SBDC- and DBDC-CV data was introduced and its usage exemplified with real datasets from the existing literature.

[11] The reason for using unadjusted data in our example is purely because the weights are not readily available.

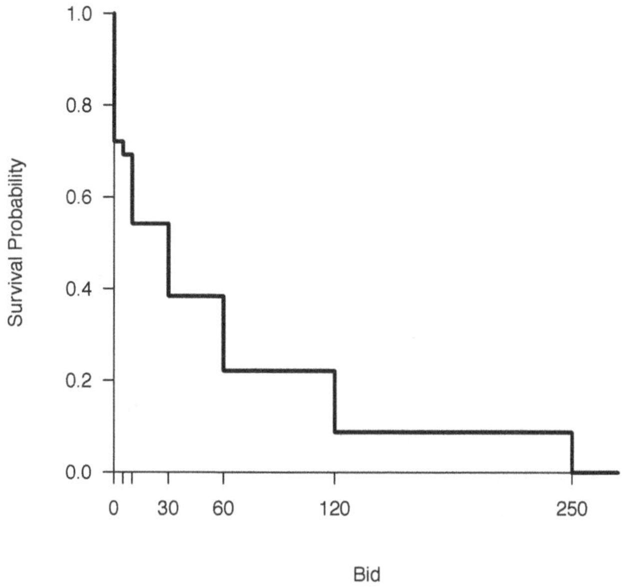

Figure 2.8 *Empirical survival function for the Exxon Valdez oil spill data (double-bounded).*

2.A Appendix

The Exxon Valdez Oil Spill data

We have `CarsonDB`, which contains a frequency table for the DBDC survey. A new data frame containing individual observations is created from the frequency table by the following steps.

The first step is to create a new data frame `db.data` to save the reconstructed individual observations and to prepare indexes:

```
R> nobs <- sum(CarsonDB[, 4:7])
R> db.data <- data.frame(bid1 = numeric(nobs),
        bid2 = numeric(nobs),
        R1 = numeric(nobs), R2 = numeric(nobs),
        bid2U = numeric(nobs), bid2L = numeric(nobs))
R> n <- rowSums(CarsonDB[, 4:7])
R> id2 <- cumsum(n)
R> id1 <- c(1, id2[1:3] + 1)
```

The second step organizes the three columns containing the first bids (`bid1`), and the increased and decreased second bids (`bid2U` and `bid2L`), respectively.

```
R> db.data$bid1[id1[1]:id2[1]] <- CarsonDB[1, 1]
R> db.data$bid1[id1[2]:id2[2]] <- CarsonDB[2, 1]
R> db.data$bid1[id1[3]:id2[3]] <- CarsonDB[3, 1]
R> db.data$bid1[id1[4]:id2[4]] <- CarsonDB[4, 1]
R> db.data$bid2U[id1[1]:id2[1]] <- CarsonDB[1, 2]
R> db.data$bid2U[id1[2]:id2[2]] <- CarsonDB[2, 2]
R> db.data$bid2U[id1[3]:id2[3]] <- CarsonDB[3, 2]
R> db.data$bid2U[id1[4]:id2[4]] <- CarsonDB[4, 2]
R> db.data$bid2L[id1[1]:id2[1]] <- CarsonDB[1, 3]
R> db.data$bid2L[id1[2]:id2[2]] <- CarsonDB[2, 3]
R> db.data$bid2L[id1[3]:id2[3]] <- CarsonDB[3, 3]
R> db.data$bid2L[id1[4]:id2[4]] <- CarsonDB[4, 3]
```

In the next step, the numbers of respondents accepting the first bids are counted bid by bid (`y1` through `y4`). Similarly, the numbers of respondents with "yy" or "ny" (`z1` through `w4`) are defined for each bid.

```
R> y1 <- CarsonDB$yy[CarsonDB$T1 == 10]
        + CarsonDB$yn[CarsonDB$T1 == 10]
R> y2 <- CarsonDB$yy[CarsonDB$T1 == 30]
        + CarsonDB$yn[CarsonDB$T1 == 30]
R> y3 <- CarsonDB$yy[CarsonDB$T1 == 60]
        + CarsonDB$yn[CarsonDB$T1 == 60]
R> y4 <- CarsonDB$yy[CarsonDB$T1 == 120]
        + CarsonDB$yn[CarsonDB$T1 == 120]
R> z1 <- CarsonDB$yy[CarsonDB$T1 == 10]
R> z2 <- CarsonDB$yy[CarsonDB$T1 == 30]
R> z3 <- CarsonDB$yy[CarsonDB$T1 == 60]
R> z4 <- CarsonDB$yy[CarsonDB$T1 == 120]
R> w1 <- CarsonDB$ny[CarsonDB$T1 == 10]
R> w2 <- CarsonDB$ny[CarsonDB$T1 == 30]
R> w3 <- CarsonDB$ny[CarsonDB$T1 == 60]
R> w4 <- CarsonDB$ny[CarsonDB$T1 == 120]
```

Using the information about the numbers of respondents calculated in the preceding step, the responses to the first and second bid variables are created.

```
R> db.data$R1[db.data$bid1 == 10][1:y1] <- 1
R> db.data$R1[db.data$bid1 == 30][1:y2] <- 1
R> db.data$R1[db.data$bid1 == 60][1:y3] <- 1
R> db.data$R1[db.data$bid1 == 120][1:y4] <- 1
R> db.data$R2[db.data$R1 == 1 & db.data$bid1 == 10][1:z1] <- 1
R> db.data$R2[db.data$R1 == 1 & db.data$bid1 == 30][1:z2] <- 1
R> db.data$R2[db.data$R1 == 1 & db.data$bid1 == 60][1:z3] <- 1
R> db.data$R2[db.data$R1 == 1 & db.data$bid1 == 120][1:z4] <- 1
```

```
R> db.data$R2[db.data$R1 == 0 & db.data$bid1 == 10][1:w1] <- 1
R> db.data$R2[db.data$R1 == 0 & db.data$bid1 == 30][1:w2] <- 1
R> db.data$R2[db.data$R1 == 0 & db.data$bid1 == 60][1:w3] <- 1
R> db.data$R2[db.data$R1 == 0 & db.data$bid1 == 120][1:w4] <- 1
```

Finally, a vector with the actual second bids (`bid2`) is created using the responses to the first bid (`R1`).

```
R> db.data$bid2[db.data$R1 == 1] <- db.data$bid2U[db.data$R1 == 1]
R> db.data$bid2[db.data$R1 == 0] <- db.data$bid2L[db.data$R1 == 0]
```

The data frame `db.data` now contains all the necessary values for the Kaplan–Meier–Turnbull estimation of the empirical survival function for the respondents' WTP.

Chapter 3

Discrete Choice Experiments

Abstract

Chapter 3 explains how to implement basic discrete choice experiments (DCEs) in R. DCEs can be applied in situations where an individual selects one alternative from a set of alternatives, each of which is expressed by a bundle of attributes, thereby revealing important attributes/levels affecting the individual's selection. Such situations occur frequently in our daily lives and therefore, DCEs have been applied in many disciplines. After defining the basic terms of DCEs, Section 3.2 presents the three steps for implementing DCEs: examining a hypothetical situation and creating a DCE design; preparing and conducting a questionnaire survey; and preparing a dataset and conducting the analysis. Section 3.3 introduces R functions to carry out these methods. In Section 3.4, three examples are given showing how the R functions can be used: two consumer valuations of rice and pork, and a residents' evaluation of a rural environment conservation plan.

3.1 Introduction

Discrete choice experiments (DCEs) have become one of the most important survey methods used in studies across a wide variety of research areas in the social sciences.[1] They can be applied in choice situations where an individual selects one alternative from a set of alternatives, each of which is expressed by a bundle of attributes, thereby revealing important attributes/levels affecting the individual's selection. As these situations occur frequently in our daily lives, DCEs have been applied in many disciplines[2]:

Food Choice situations are generally constructed in such a way that individuals are requested to select their most preferred alternative from two or more items of a specific food (e.g., oranges, beef, or wine). Then, their valuations

[1]This chapter is a revised version of Aizaki (2012a).

[2]The applications presented in the text are not comprehensive examples. For more detailed information, readers should refer to books on DCEs and/or review papers that specialize in a particular discipline, such as Bennett and Blamey (2001), Rolfe and Bennett (2006), Birol and Koundouri (2008), Bennett and Birol (2010), and de Bekker-Grob et al. (2012).

of the characteristics of the food are measured in terms of a monetary unit (willingness to pay). Besides price, these characteristics include:

- The food's ordinal characteristics: appearance (Alfnes et al., 2006), country of origin (Unterschultz et al., 1998), and brand names (Jaeger and Rose, 2008).
- Food safety: hormones (Alfnes, 2004), genetic modification technology (Ding et al., 2012), and animal cloning (Lusk and Marette, 2010).
- Food-related values: animal welfare (Lagerkvist et al., 2006), fair trade (Onozaka and Mcfadden, 2011), and environmental issues (Lusk et al., 2007).

Environment DCEs measure the economic values of various resources and resource-related products/services by presenting individuals with a trade-off between the benefits and costs related to the resource. For example:

- Natural resource management: wildlife (Adamowicz et al., 1998), water (Scarpa et al., 2007), and land (Johnston and Duke, 2007).
- Recreational activities: water-based activities (Adamowicz et al., 1994), climbing (Hanley et al., 2002), and hunting (Boxall et al., 1996).
- Products/services: energy (Roe et al., 2001), fuels (Susaeta et al., 2010), and recycling services (Karousakis and Birol, 2008).

Medical and health care DCE applications in this field capture the preferences of patients and professionals related to the following examples and then evaluate these in terms of a monetary or non-monetary measure:

- Tests and treatments: screening tests (Gerard et al., 2003), surgery (McIntosh and Ryan, 2002), and vaccinations (Hall et al., 2002).
- Quality of life: social care outcomes (Ryan et al., 2006), health outcomes (Burr et al., 2007), and priority of a patients' waiting list (Witt et al., 2009).
- Products/services: cigarettes (Goto et al., 2007), drugs (Dickie and Messman, 2004), and hip protectors (Telser and Zweifel, 2002).

Transportation By imitating various transport-related choice situations, DCEs have been used to examine factors affecting choice behavior, evaluate travel time, or predict market share of transport modes, for example:

- Route choice: car drivers (Hensher and Sullivan, 2003), freight transporters (Puckett and Hensher, 2009), and air travelers (Bliemer and Rose, 2011).
- Transport mode choice: commuters (Greene et al., 2006), travelers (Molin and Timmermans, 2010), and companies (Masiero and Hensher, 2010).
- Location choice: houses (Molin et al., 1996), companies (Leitham et al., 2000), and convention sites (Crouch and Louviere, 2003).

To help us understand DCEs, several studies have discussed the various ways in which DCEs can be applied (Louviere et al., 2000; Bennett and Blamey, 2001; Bateman et al., 2002; Hensher et al., 2005; Kanninen, 2007; Ryan et al., 2008a). However, it may not be easy for those who are relatively unfamiliar with DCEs to create a DCE design and conduct a statistical analysis of the responses to DCE questions without additional support. This chapter explains how to implement a basic DCE in R for DCE beginners, focusing on creating choice sets using orthogonal main-effect designs (OMEDs) and analyzing respondents' choice behaviors in the choice sets using a conditional logit (CL) or binary logit (BL) model.

Four packages are mainly used in this chapter: **support.CEs** (Aizaki, 2014b), **survival** (Therneau, 2014), **stats** (R Core Team, 2014), and **mded** (Aizaki, 2012b). The **support.CEs** package provides seven basic functions for DCEs: two functions to create DCE designs based on OMEDs; a function to convert a DCE design into questionnaire format; a function to convert a DCE design into a design matrix; a function that creates a dataset suitable for statistical analysis of the responses to the DCE questions; and two functions to calculate the goodness-of-fit measures of an estimated model and the marginal willingness to pay (MWTP) for the attribute levels of the estimated model, respectively. The `clogit()` function in the **survival** package and the `glm()` function in the **stats** package are used in the statistical analysis of responses to DCE questions. The `mded()` function in the **mded** package measures the difference between two independent or non-independent empirical distributions such as the MWTP.

After providing an overview of DCEs, Section 3.2 presents the steps for implementing DCEs, while Section 3.3 introduces the R functions used to carry out these methods. Section 3.4 provides three examples showing how these R functions can be used: two consumer valuations of rice and pork, respectively, and a residents' valuation of a rural environment conservation plan.

3.2 Overview of DCEs

3.2.1 Basic terms used in DCEs

Before giving an overview of the method for implementing DCEs, we define the terms used in this chapter. An "attribute" is a characteristic or feature of an alternative, while a "level" or "attribute level" represents the numerical or qualitative value of the attribute in a given alternative. An "alternative" is a combination of attributes; that is, one alternative can have two or more attributes. A "choice set" refers to the set of alternatives available to individuals. A single choice set includes two or more alternatives with at least one attribute having different levels between alternatives, and possibly including an opt-out alternative. In a DCE, respondents are usually asked to select their most preferred alternative from a choice set; therefore, one choice set constitutes a DCE question (choice situation). A "DCE design" refers to the collection of individual choice sets.

Figure 3.1 shows an example of a DCE design comprising a total of nine choice sets (Q1 to Q9). Each choice set consists of three alternatives ("Alternative 1," "Alternative 2," and an opt-out option). "Alternative 1" and "Alternative 2" each consist of three attributes: attribute A with the three levels

Q1. Please select your most preferred alternative from the following:

	Alternative 1	Alternative 2
Attribute A	a2	a3
Attribute B	b2	b3
Attribute C	c2	c3

[ ] I select alternative 1.
[ ] I select alternative 2.
[ ] I do not select either of these.

Q2. Please select your most preferred alternative from the following:

	Alternative 1	Alternative 2
Attribute A	a2	a1
Attribute B	b3	b2
Attribute C	c1	c3

[ ] I select alternative 1.
[ ] I select alternative 2.
[ ] I do not select either of these.

...

Q9. Please select your most preferred alternative from the following:

	Alternative 1	Alternative 2
Attribute A	a3	a2
Attribute B	b2	b2
Attribute C	c1	c3

[ ] I select alternative 1.
[ ] I select alternative 2.
[ ] I do not select either of these.

Figure 3.1 *An example of a DCE design.*

Q1. Would you like to purchase this tomato?

	Tomato
Region of origin	Region A
Eco-friendly	Most
Price	$1.2

[ ] Yes.
[ ] No.

Figure 3.2 *Example of a BDCE with an opt-out option.*

"a1," "a2," and "a3"; attribute B with the three levels "b1," "b2," and "b3"; and attribute C with the three levels "c1," "c2," and "c3." In Q1, the characteristics of "Alternative 1" are "a2," "b2," and "c2," whereas those of "Alternative 2" are "a3," "b3," and "c3." Respondents are requested to select their most preferred alternative from the three options including the opt-out option. The combinations of attribute levels are changed for each question, with the questions repeated nine times in the example.

DCEs can be categorized from various viewpoints. As an initial viewpoint, DCEs can be divided into two subcategories with respect to the number of alternatives per question: multinomial DCEs, in which respondents are requested to select one of three or more alternatives included in each question, and binary DCEs, in which respondents are requested to select one of only two alternatives included in each question. This classification is similar to that used in discrete choice models, i.e., multinomial choice models (e.g., CL model) and binary choice models (e.g., BL model). Multinomial DCEs have been widely applied in a variety of fields, whereas binary DCEs seem to be more frequently used in health and medical economics research. In the following, the term "BDCE" is used when it is necessary to refer explicitly to binary DCEs.

As a second viewpoint, DCEs can also be divided into three subcategories with respect to the type of alternatives in the choice sets. The first type is a DCE with an opt-out option shown in Figure 3.1. This option allows respondents not to choose one of the alternatives presented and is expressed as "none of these" or "no choice." In a BDCE with an opt-out option, only one alternative is presented to respondents in each question and the respondents need to decide whether or not to select the alternative (Figure 3.2). The second type is a forced choice formatted DCE in which one of the alternatives in each question must be selected; that is, there is no opt-out option in any of the questions (Figure 3.3). The third type is a DCE with a common base option: the same option appears in all choice sets (Figure 3.4). Such an option could

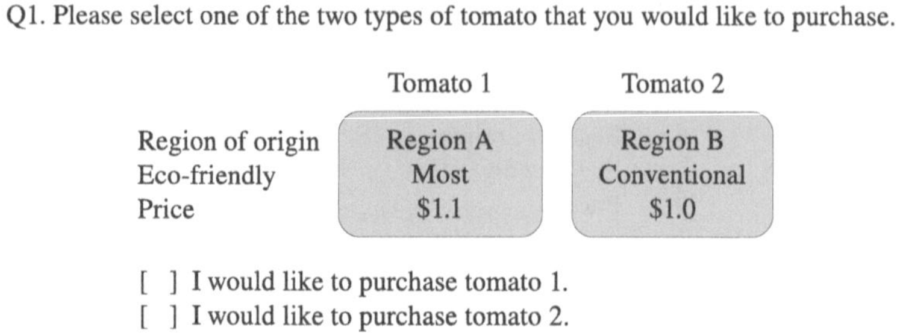

Figure 3.3 *Example of a forced choice formatted BDCE.*

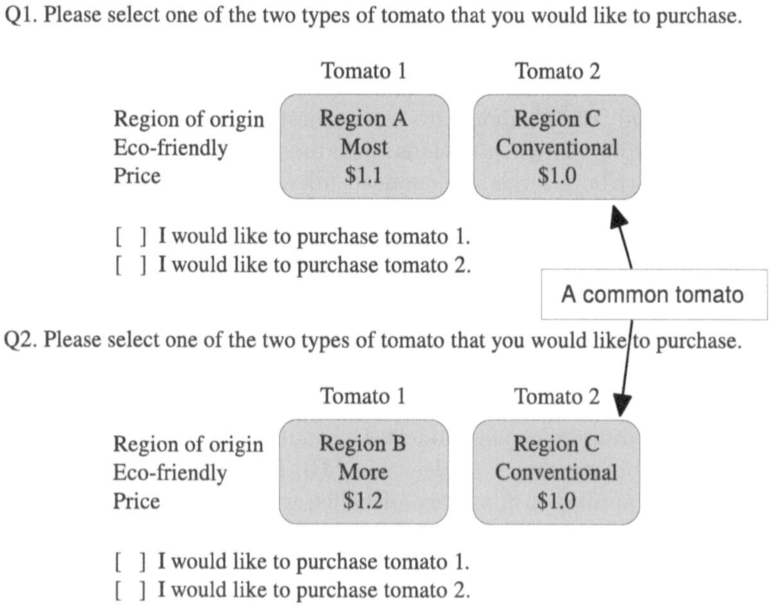

Figure 3.4 *Example of a BDCE with a common base option.*

be, for example, the status quo, a predicted situation in which no policy will be implemented, or the standard product alternative.[3]

DCEs can further be divided into two subcategories with respect to the

[3]Because BDCEs may be unfamiliar in some fields, example papers applying BDCEs are listed below according to the various categories. BDCEs with an opt-out option have been used by Telser and Zweifel (2002), Goto et al. (2007), and Tappenden et al. (2007). Forced choice BDCEs have been employed by Farrar et al. (2000), Torbica and Fattore (2010), Kolstad (2011), and Laver et al. (2011), while BDCEs with a common base option have been applied by Ryan (1999), Longworth et al. (2001), Ida and Goto (2009), and McNair et al. (2011).

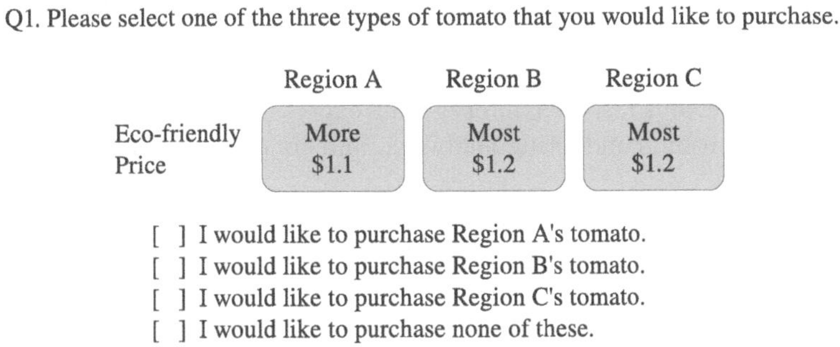

Figure 3.5 *A DCE question in a labeled design example.*

labeling of alternatives: a labeled type and unlabeled (generic) type. In labeled DCEs, the alternatives are named (labeled) so that each alternative in a choice set can be distinguished from the others. For example, the first alternative is labeled "Region A's tomato," the second "Region B's tomato," and the third "Region C's tomato" (Figure 3.5). In an unlabeled DCE, each alternative is given a generic name; for example, the first alternative is named "Alternative 1" and the second "Alternative 2" (Figure 3.1). In a labeled DCE, the attribute can be an alternative-specific attribute, which refers to one which is included in only one alternative; in an unlabeled DCE, the attribute is a generic attribute, which is included in all the alternatives.

3.2.2 Steps for implementing DCEs

This subsection explains the DCE methods implemented using R functions in Sections 3.3 and 3.4. For a comprehensive and detailed explanation of DCE methods, the reader is referred to Bateman et al. (2002), Bennett and Blamey (2001), Hensher et al. (2005), Holmes and Adamowicz (2003), Kanninen (2007), Lancsar and Louviere (2008), Louviere et al. (2000), and Ryan et al. (2008a), amongst others. In particular, Lancsar and Louviere (2008) provides a useful checklist for implementing DCEs.

After establishing the objectives of applying DCEs, DCE implementation can roughly be summarized in three steps: (1) examining a hypothetical situation and creating a DCE design; (2) preparing and conducting a questionnaire survey; and (3) preparing a dataset and conducting an analysis.

Step 1: Examining a hypothetical situation and creating a DCE design

As a first step, the DCE user must consider a hypothetical choice situation in which respondents are requested to answer DCE questions. Considerations in describing this situation include: (1) characterizing the situation in which respondents need to make a decision; (2) specifying the alternatives from which the respondents must select; (3) designating attributes of the alternatives and attribute levels; (4) deciding whether an opt-out option is included in the choice set; and (5) selecting the number of alternatives per choice set. These considerations are implemented through one or more of the following methods: reviewing previous research, interviewing experts/people involved in the hypothetical situation, conducting fieldwork, and conducting pilot surveys and/or focus groups.[4]

After defining the hypothetical situation, a DCE design is created accordingly. Here we introduce three methods for creating choice sets based on OMEDs as explained by Johnson et al. (2007) (see also Chrzan and Orme, 2000, and associated references) including a rotation design method, mix-and-match method, and L^{MA} method.[5] Whereas the rotation and mix-and-match methods create an unlabeled type of DCE design, which can contain generic attributes in the utility function, the L^{MA} method creates a labeled type of DCE design, which can contain both generic and alternative-specific attributes. These methods can be implemented using the R functions explained in the next section.[6]

The rotation design method uses an OMED as the first alternative in each choice set; this method creates one or more additional alternative(s) by adding a constant to each attribute level of the first alternative; the k-th (≥ 2) alternative in the j-th ($= 1, 2, ..., J$) choice set is created by adding one to each of the m attributes in the $(k-1)$-th alternative in the j-th choice set. If the level of the attribute in the $(k-1)$-th alternative is the maximum, then the level of the attribute in the k-th alternative is assigned the minimum value.

The mix-and-match method modifies the rotation method by introducing a randomizing process. After placing a set of N alternatives created from an OMED into an urn, one or more additional set(s) of N alternatives is/are created using the rotation method and placed into different urn(s). A choice set is generated by randomly selecting one alternative from each urn. This selection process is repeated, without replacement, until all the alternatives are assigned to N choice sets. These N choice sets correspond to the DCE design.

The L^{MA} design method creates a DCE design directly from the OMED. In this method, an OMED with M times A columns of L level factors is used

[4]Coast and Horrocks (2007) present qualitative methods for developing attributes and their levels based on a health economics case study.

[5]Although a review of the development of the theory and practice of designing choice sets for DCEs is beyond the scope of this book, the process of designing choice sets is a crucial aspect of conducting a DCE survey. Readers who are interested in DCE designs should refer to previous works, such as those by Louviere et al. (2000), Kanninen (2007), Ferrini and Scarpa (2007), Street and Burgess (2007), Ryan et al. (2008a), Scarpa and Rose (2008), Bliemer and Rose (2011), and Johnson et al. (2013).

[6]Studies in which the rotation design method has been applied include Burr et al. (2007), Carlsson et al. (2007), de Bekker-Grob et al. (2008), Asrat et al. (2010), and de Bekker-Grob et al. (2010), while those applying the mix-and-match design method include works by Enneking (2004) and Enneking et al. (2007). The L^{MA} design method is employed in works by Morrison et al. (2002), Kallas et al. (2007), Shapansky et al. (2008), Brooks and Lusk (2010), Lusk and Marette (2010), and Pek and Jamal (2011).

to create each choice set containing M alternatives of A attributes with L levels. Each row in the design corresponds to the alternatives in a choice set.

It should be noted that application of an OMED to create choice sets is a restrictive approach in which all main effects can be estimated, but two-way or higher-order interaction effects cannot. If interaction effects are deemed to be important in determining respondent choice, an alternative design approach should be considered (e.g., Street and Burgess, 2007). It should also be noted that interaction effects can have an influence on main effects.

When a large DCE design is created (that is, a DCE with numerous questions), the respondent may find it psychologically demanding to answer all the questions. In such cases, the DCE design is frequently divided into two or more blocks (subsets) of choice sets (questions), and each respondent is asked to answer only a single block of questions. To determine the block size, it is necessary to set the number of questions per respondent. For example, de Bekker-Grob et al. (2010) reviewed 114 DCEs studies in health economics and revealed that 39% of the studies asked each respondent eight or fewer questions, 38% posed 9–16 questions per respondent, and 18% asked each respondent more than 16 questions. Almost all of the 15 DCE case studies in developing countries presented in Bennett and Birol (2010) used eight or fewer DCE questions per respondent. However, there is another view regarding the number of questions per respondent: Lancsar and Louviere (2008) pointed out that respondents may be asked more questions than would normally be considered appropriate.

Step 2: Preparing and conducting a questionnaire survey

The second step in implementing a DCE is to prepare and conduct a questionnaire survey. This step includes the following sub-steps: writing questions (DCE and other questions); creating a questionnaire; determining a sampling frame corresponding to the targeted population; selecting a survey mode; deciding on a sampling method; choosing a sample size; and sampling the respondents from the sampling frame.

Although these sub-steps are similar to those employed in general questionnaire surveys (e.g., Mangione, 1995; Dillman, 2000),[7] it should be noted that the sample size may differ from the number of respondents. When each respondent is asked to answer multiple questions, the sample size is calculated by multiplying the number of respondents by the number of questions per respondent.

Thus far, there is no established method for determining the optimal number of respondents that could affect the quality of a DCE study; however, Louviere et al. (2000) and Hensher et al. (2005) recommended using an equa-

[7] It is necessary to eliminate various factors that bias DCE estimation results from questions. See the comprehensive and detailed works mentioned above for DCE survey techniques. Guidelines for contingent valuation methods are also valuable for designing DCE surveys (e.g., Mitchell and Carson, 1989).

tion to calculate the minimum number of random samples that should be collected for an arbitrary number of DCE questions per respondent.[8]

Step 3: Preparing a dataset and conducting the analysis

As the final step in implementing a DCE, the user has to prepare a dataset from the valid samples and conduct a statistical analysis based on the dataset. Because respondents in a DCE question are asked to select their most preferred alternative from a set of alternatives, the structure of the dataset for DCEs is different from that usually found in statistical analysis. In addition, the structure often differs according to the software package used for DCE analysis. Therefore, the way a dataset is created in this chapter is explained in the next section.

As discussed in Chapter 1, respondents' decision-making can be modeled under random utility theory and thus can be analyzed using discrete choice models. Here we use the CL model; the probability of respondent n's selection of alternative i from choice set S is modeled as:

$$P_n(i) = \frac{\exp(V_{in})}{\sum_{j \in S} \exp(V_{jn})}, \tag{3.1}$$

where V_{in} is respondent n's systematic component of utility for alternative i. This equation is the same as Eq. (1.5).

V_{in} for an unlabeled DCE, assumed to be a linear additive function of the independent variables X_{ikn}, is defined as:

$$V_{in} = \alpha + \sum_{k=1}^{K-1} \beta_k X_{ikn}, \tag{3.2}$$

where α is a constant and variable X_{ikn} refers to a generic variable, that is, X_{ikn} has a common (generic) coefficient β_k for alternative i.

If the DCE is a labeled type, Eq. (3.2) is rewritten as:

$$V_{in} = \alpha_i + \sum_{k=1}^{K-1} \beta_{ik} X_{ikn}, \tag{3.3}$$

where α_i is an alternative-specific constant (ASC) and β_{ik} is the coefficient of variable X_{ikn}. In Eq. (3.3), β_{ik} in alternative i can take a value that is different from that in alternative j (β_{jk}); in other words, the variables refer to alternative-specific variables. When applying a labeled DCE, it may be feasible to test whether the coefficients are generic or alternative-specific (see Section 3.4.2).

A constant/ASC is not included in an arbitrary alternative. For example,

[8]Hensher et al. (2005) pointed out that the equation for calculating the minimum random sample size is mistyped in Louviere et al. (2000); see Hensher et al. (2005) for the correct equation.

in the case of a DCE with an opt-out option, all V_i components, except for the opt-out option, usually have constants/ASCs.

If there are only two alternatives i and j in the choice set (that is, a BDCE), Eq. (3.1) is rewritten as:

$$P_n(i) = \frac{1}{1 + \exp(V_{jn} - V_{in})}. \tag{3.4}$$

This is a BL model and is the same as Eq. (1.6).

After estimating a model using the maximum likelihood estimation procedure, testing the estimates, and evaluating the estimated model as explained in Chapter 1, we can derive welfare measures from the estimated model that are widely used in studies in which DCEs are applied.[9] Under Eq. (3.2), the MWTP for a non-monetary variable X_N is the economic value of a small change in X_N calculated as:

$$MWTP = -\frac{\partial V/\partial X_N}{\partial V/\partial X_M} = -\frac{\beta_N}{\beta_M}, \tag{3.5}$$

where X_M is the monetary variable, and β_N and β_M are coefficients of X_N and X_M, respectively.

If multiple alternatives are available, a change in the attributes of the alternatives affects the choice probability of each alternative. The compensating variation (C)[10] of a change under Eq. (3.2) is calculated as:

$$C = -\frac{1}{\lambda}[\ln \sum_{i \in S} \exp(V_i^0) - \ln \sum_{i \in S} \exp(V_i^1)], \tag{3.6}$$

where V_i^0 and V_i^1 are the values of V_i before and after the change, respectively, and λ is the marginal utility of income. However, λ usually refers to a negative value of the coefficient of the monetary variable.

If only a single alternative is available before and after the change, Eq. (3.6) is rewritten as

$$C = -\frac{1}{\lambda}(V_i^0 - V_i^1). \tag{3.7}$$

This situation is referred to as the "state of the world" (Bennett and Adamowicz, 2001).

Furthermore, confidence intervals for the welfare measures and the differences between two independently estimated welfare measures are frequently calculated. See Chapter 1 for the details of these calculation methods.

---

[9] For a comprehensive explanation and calculation examples of welfare measures in discrete choice models, see for example, Bennett and Adamowicz (2001), Bennett et al. (2001), Bockstael and Freeman (2005), Amaya-Amaya et al. (2008), Ryan et al. (2008b), and Louviere and Fiebig (2010).

[10] Regardless of the difference between the compensating variation and the compensating surplus, the former term is used according to Bockstael and Freeman (2005).

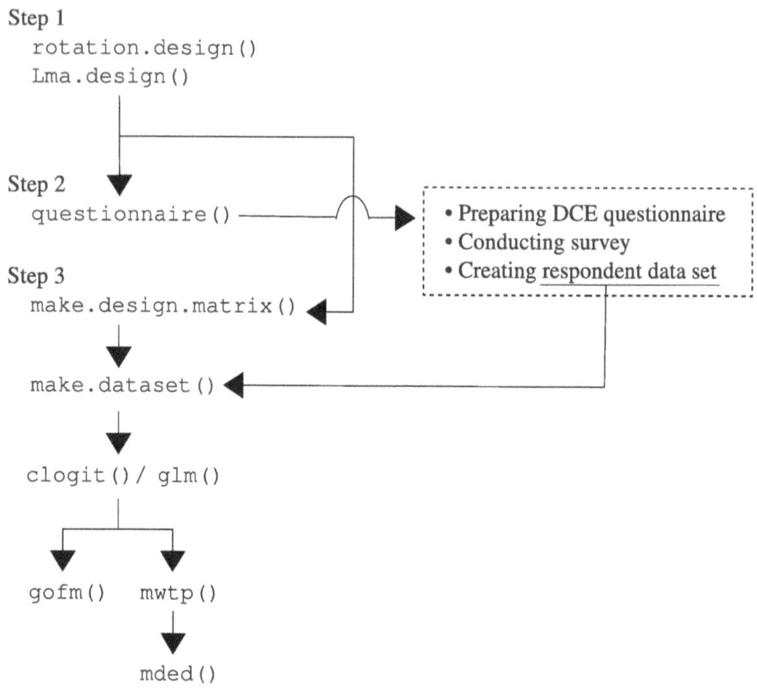

Figure 3.6 *Functions used in each step of implementing a DCE.*

3.3 R functions for DCEs

3.3.1 Overview

Figure 3.6 illustrates the functions used in each step of implementing a DCE. In the first step, two functions are available: `rotation.design()` and `Lma.design()` from the **support.CEs** package. The `rotation.design()` function implements a rotation design method and a mix-and-match design method, while the `Lma.design()` function realizes the L^{MA} design method. Each function depends on an OMED generated by the `oa.design()` function in the **DoE.base** package (Grömping, 2014b).[11]

In the second step, the `questionnaire()` function in package **support.CEs** is available to convert the DCE design created by `rotation.design()` or `Lma.design()` into DCE questions. A questionnaire for the DCE study is prepared using the created DCE questions and other questions related

[11]Users should check the design generated by `oa.design()` to confirm whether it suits their requirements. For example, if the function does not find an array corresponding to these requirements, it returns a full factorial. In addition, the function contains limited arrays. The reader should refer to the relevant reference manual for the detailed features of this function.

to the respondents' individual characteristics such as age, gender, attitudes, knowledge, and opinions. After conducting the survey, the respondent dataset, which usually includes the responses to both the DCE questions and the other questions, is created by the user.

In the last step, CL model analysis of the responses to the DCE questions can be carried out using the `clogit()` function in the **survival** package. When using this function to analyze the responses to the DCE questions, a dataset in a special format is needed: each alternative should comprise one row of this dataset. To create such a dataset, two functions in **support.CEs** can be used. The first is the `make.design.matrix()` function, which converts a DCE design created by `rotation.design()` or `Lma.design()` into a design matrix, where each alternative comprises one row of the matrix. The second is the `make.dataset()` function, which creates a dataset for use by `clogit()` by combining the dataset with information about the responses to the DCE questions and that containing the design matrix related to these questions created by `make.design.matrix()`.

In the case of a BDCE, BL model analysis can be conducted using the `glm()` function in the **stats** package. Functions `make.design.matrix()` and `make.dataset()` can similarly be applied to construct a dataset suitable for analysis using `glm()`.

The `gofm()` function in **support.CEs** provides ρ^2 and $\bar{\rho}^2$. This function also displays the number of estimated coefficients, the log-likelihood values at the start and at convergence, and the information criteria. The `mwtp()` function in **support.CEs** gives the MWTP in the case of a simple linear utility function from the model estimates and confidence interval for the MWTP. Finally, the `mded()` function in the **mded** package calculates the difference between the independently estimated MWTP values and conducts a statistical test on the difference according to the complete combinatorial method.

In preparation for implementing the example code using the functions given in Sections 3.3 and 3.4, we start by loading the three packages into R:

```
R> library(support.CEs)
R> library(survival)
R> library(mded)
```

3.3.2 Creating a DCE design

The `rotation.design()` function corresponds to the rotation design method and mix-and-match method, while the `Lma.design()` function corresponds to the L^{MA} design method. A generic call to each of these functions is given below:

```
rotation.design(candidate.array = NULL,
                attribute.names,
                nalternatives,
```

```
                nblocks,
                row.renames = TRUE,
                randomize = FALSE,
                seed = NULL)

Lma.design(candidate.array = NULL,
           attribute.names,
           nalternatives,
           nblocks,
           row.renames = TRUE,
           seed = NULL)
```

The two functions have six common arguments:

- `candidate.array`: A data frame containing an array created by the user. This argument usually does not need to be set by the user.
- `attribute.names`: A list of attributes and levels.
- `nalternatives`: An integer variable containing the number of alternatives per choice set, excluding an opt-out or common base option.
- `nblocks`: An integer variable containing the number of blocks into which the design is divided.
- `row.renames`: A logical variable, which denotes, if `TRUE` that integer values are assigned to the row names starting from `1`, and if `FALSE` that the row names are not changed.
- `seed`: Seed for a random number generator.

An argument that is only valid for function `rotation.design()` is:

- `randomize`: A logical variable, which denotes, if `TRUE` that the mix-and-match method is executed, and if `FALSE` that the rotation method is executed.

The `Lma.design()` function can also be used to create a BDCE design based on an OMED by setting the argument `nalternatives` to `1` for a BDCE with an opt-out or common base option or `2` for a forced choice format of the BDCE.

Two points should be noted with regard to setting arguments. First, argument `nblocks` in `rotation.design()` must be a divisor of the number of choice sets included in the resultant DCE design. Whereas `Lma.design()` divides choice sets into sub-blocks based on a factor with `nblocks` levels, `rotation.design()` randomly divides choice sets into sub-blocks based on `nblocks`. An error message is displayed if this condition is not met.

Second, it is possible that these functions do not find a design that satisfies the user's requirements. To create a DCE design under default settings, the two functions use an OMED that is internally produced by `oa.design()` based on argument `attribute.names`. However, if there is no OMED corresponding to `attribute.names`, `oa.design()` returns a full factorial based on `attribute.names`. If these functions cannot create a DCE design matching the user's requirements, the user can still obtain a design by assigning an arbitrary (user-defined) array to argument `candidate.array`; the functions then use this array to create a DCE design. When a user-defined array is used, the last column of the array must contain a column to divide the design based on `nblocks`. Arguments `attribute.names` and `nblocks` must also be assigned according to this array.

Example code and output

As an example, choice sets are created using the mix-and-match method according to the following conditions:

- A product alternative has three attributes, each of which has three levels: attribute X with the three levels "x1," "x2," and "x3"; attribute Y with the three levels "y1," "y2," and "y3"; and attribute Z with the three levels "10," "20," and "30."
- Two product alternatives and the opt-out option appear in each choice set.
- The choice sets are divided into 3 blocks.
- The random generator seed is set to 123.

Code corresponding to the above and the output thereof are given below:

```
R> rd <- rotation.design(
            attribute.names = list(X = c("x1", "x2", "x3"),
                                   Y = c("y1", "y2", "y3"),
                                   Z = c("10", "20", "30")),
            nalternatives = 2,
            nblocks = 3,
            seed = 123)
The columns of the array have been used in order of appearance.
For designs with relatively few columns,
the properties can sometimes be substantially improved
using option columns with min3 or even min34.

R> rd
Choice sets:
alternative 1 in each choice set
  BLOCK QES ALT  X  Y  Z
1     1   1   1 x1 y1 10
2     1   2   1 x1 y3 20
3     1   3   1 x3 y3 30
4     2   1   1 x1 y2 30
5     2   2   1 x2 y1 30
6     2   3   1 x3 y1 20
7     3   1   1 x2 y3 10
8     3   2   1 x3 y2 10
9     3   3   1 x2 y2 20
```

```
alternative 2 in each choice set
  BLOCK QES ALT  X  Y  Z
1     1   1   2 x2 y2 20
2     1   2   2 x2 y1 30
3     1   3   2 x1 y1 10
4     2   1   2 x2 y3 10
5     2   2   2 x3 y2 10
6     2   3   2 x1 y2 30
7     3   1   2 x3 y1 20
8     3   2   2 x1 y3 20
9     3   3   2 x3 y3 30

Candidate design:
  A B C
1 1 3 2
2 3 1 2
3 3 3 3
4 2 3 1
5 2 2 2
6 1 1 1
7 1 2 3
8 3 2 1
9 2 1 3
class=design, type= oa

Design information:
number of blocks = 3
number of questions per block = 3
number of alternatives per choice set = 2
number of attributes per alternative = 3
```

Each design function returns an object of the S3 class "`cedes`". Because a generic function `print()` for the class is defined, the output displayed in the above example is formatted in a manner that is easy to comprehend.

The object in the class consists of three subcomponents. The first is a DCE design that is provided as a list (`alternatives`) of the j-th alternative (`alt.j`) in each choice set. Each `alt.j` contains attribute variables corresponding to argument `attribute.names`, variable `BLOCK` denoting the serial number of the blocks, variable `QES` denoting the serial number of the DCE questions for each value of `BLOCK`, and variable `ALT` denoting the serial number of the alternatives for each value of `QES`. The second is the candidate set (`candidate`), which is either generated by `oa.design()` or set by the user for `candidate.array`. The DCE design is created using the candidate set in conjunction with the design method selected by the user (i.e., rotation, mix-and-match, or L^{MA} method). The third is information related to the DCE design (`design.information`):

the number of blocks into which the DCE design is divided; the number of questions per block; the number of alternatives per choice set, excluding any opt-out or common base option; and the number of attributes per alternative.

Individual components in the output are accessible using the $ operator. For example, the following code returns the first alternative in `rd`:

```
R> rd$alternatives$alt.1
  BLOCK QES ALT  X  Y  Z
1     1   1   1 x1 y1 10
2     1   2   1 x1 y3 20
3     1   3   1 x3 y3 30
4     2   1   1 x1 y2 30
5     2   2   1 x2 y1 30
6     2   3   1 x3 y1 20
7     3   1   1 x2 y3 10
8     3   2   1 x3 y2 10
9     3   3   1 x2 y2 20
```

As seen in the above example, messages are frequently displayed immediately after executing one of the two functions if the function works correctly. These messages are taken from `oa.design()` and may be valuable to a user wishing to define the original array and assign it to `candidate.array`. For a detailed explanation of these messages, refer to the help for `oa.design()`.

See Sections 3.4.2 and 3.4.3 for examples of creating DCE and BDCE designs, respectively, using `Lma.design()`.

3.3.3 *Converting a DCE design into questionnaire format*

The `questionnaire()` function converts a DCE design created by one of the functions `rotation.design()` or `Lma.design()` into typical DCE questions used in a questionnaire survey. A generic call to the function and an explanation of its arguments are given below:

```
questionnaire(choice.experiment.design,
              common = NULL,
              quote = TRUE)
```

- `choice.experiment.design`: A data frame containing a DCE design created by `rotation.design()` or `Lma.design()`.
- `common`: A vector containing a fixed combination of attribute levels corresponding to a common base option in each question if such an option exists.
- `quote`: A logical variable that denotes whether the attribute levels in each question are printed with or without quotation marks, depending on whether it is `TRUE` or `FALSE`, respectively.

Example code and output

For example, the DCE design `rd`, which was created in Section 3.3.2, can be converted into questions as follows:

```
R> questionnaire(choice.experiment.design = rd)
Block 1

Question 1
  alt.1 alt.2
X "x1"  "x2"
Y "y1"  "y2"
Z "10"  "20"

Question 2
  alt.1 alt.2
X "x1"  "x2"
Y "y3"  "y1"
Z "20"  "30"

Question 3
  alt.1 alt.2
X "x3"  "x1"
Y "y3"  "y1"
Z "30"  "10"

Block 2

...

Block 3

...

Question 3
  alt.1 alt.2
X "x2"  "x3"
Y "y2"  "y3"
Z "20"  "30"
```

The resultant questions for the DCE are displayed on the R console. It is thus easy to copy the questions from the R console, paste them into other software packages such as a word processor, and then modify them to create the questionnaire if necessary.

3.3.4 Creating a design matrix

The `make.design.matrix()` function converts a DCE design created by one of the functions `rotation.design()` or `Lma.design()` into a design matrix that is suitable for CL model analysis using function `clogit()` or BL model analysis using function `glm()`. A generic call to the function and definitions of its arguments are given below:

```
make.design.matrix(choice.experiment.design,
                   optout = TRUE,
                   categorical.attributes = NULL,
                   continuous.attributes = NULL,
                   unlabeled = TRUE,
                   common = NULL,
                   binary = FALSE)
```

- `choice.experiment.design`: A data frame containing a DCE design created by `rotation.design()` or `Lma.design()`.
- `optout`: A logical variable that denotes whether or not the opt-out alternative is included in the design matrix created by this function depending on its value of `TRUE` or `FALSE`, respectively.
- `categorical.attributes`: A vector containing attributes treated as categorical independent variables in the model analysis.
- `continuous.attributes`: A vector containing attributes treated as continuous independent variables in the model analysis.
- `unlabeled`: A logical variable that describes the type of the DCE design represented by `choice.experiment.design` as unlabeled or labeled, depending on its value of `TRUE` or `FALSE`, respectively.
- `common`: A vector containing a fixed combination of attribute levels corresponding to a common base option in each question. If there is no common base option, it is set to `NULL`.
- `binary`: A logical variable, which denotes whether a design matrix has been created for a BL or CL model, depending on its value of `TRUE` or `FALSE`, respectively.

Two points should be noted with regard to continuous and categorical variables in `make.design.matrix()`. First, the level of `continuous.attributes` can only take numerical values; that is, the level should not contain attribute units, such as "USD," "kg," or "km." For example, if `continuous.attributes` is set to `c("Z")` and shows the price attribute of a product alternative, variable `Z` must not contain the levels `USD10`, `USD20`, and `USD30`. Instead, the variable should be set to `10`, `20`, or `30`, respectively.

Second, categorical variables created by the function are not in factor format. R usually treats categorical variables as factors. However, the values of attribute variables in each row corresponding to an opt-out option must be set to zero because the systematic component of the utility of the opt-out option is normalized to zero. Therefore, `make.design.matrix()` converts categorical attributes into dummy variables. In other words, the minimum value in a categorical attribute is normalized; as a result, each dummy variable is equal to 1 if the categorical attribute takes on a value other than the minimum value. The dummy variables are referred to by their levels.

For example, in a categorical attribute X with three levels, namely, "x1," "x2," and "x3," dummy variables are created as follows: (1) If the DCE design is unlabeled, two dummy variables are created: a dummy variable `x2` is equal to `1` when attribute X takes "x2," and `0` otherwise; and a dummy variable `x3` is equal to `1` when attribute X takes "x3," and `0` otherwise. (2) If the DCE design is labeled and the design contains two alternatives ("alternative 1" and "alternative 2"), excluding an opt-out alternative, four dummy variables are created: a dummy variable `x21` is equal to `1` when attribute X in alternative 1 takes "x2," and `0` otherwise; a dummy variable `x22` is equal to `1` when attribute X in alternative 2 takes "x2," and `0` otherwise; a dummy variable `x31` is equal to `1` when attribute X in alternative 1 takes "x3," and `0` otherwise; and a dummy variable `x32` is equal to `1` when attribute X in alternative 2 takes "x3," and `0` otherwise.

Example code and output

As an example, we create a design matrix under the following conditions:

- An unlabeled DCE design is stored in `rd`, which was created in Section 3.3.2.
- An opt-out option is included in the choice tasks.
- Each alternative has three attributes: attributes X and Y are treated as categorical variables and attribute Z is treated as a continuous variable.

Code corresponding to the above conditions together with the output thereof is given below:

```
R> dm.rd <- make.design.matrix(
              choice.experiment.design = rd,
              optout = TRUE,
              categorical.attributes = c("X", "Y"),
              continuous.attributes = c("Z"),
              unlabeled = TRUE,
              common = NULL,
              binary = FALSE)
R> dm.rd
  BLOCK QES ALT ASC x2 x3 y2 y3  Z
1     1   1   1   1  0  0  0  0 10
2     1   1   2   1  1  0  1  0 20
3     1   1   3   0  0  0  0  0  0
4     1   2   1   1  0  0  0  1 20
5     1   2   2   1  1  0  0  0 30
6     1   2   3   0  0  0  0  0  0
```

```
7         1   3   1   1   0   1   0   1  30
8         1   3   2   1   0   0   0   0  10
9         1   3   3   0   0   0   0   0   0
10        2   1   1   1   0   0   1   0  30
11        2   1   2   1   1   0   0   1  10
12        2   1   3   0   0   0   0   0   0
13        2   2   1   1   1   0   0   0  30
14        2   2   2   1   0   1   1   0  10
15        2   2   3   0   0   0   0   0   0
16        2   3   1   1   0   1   0   0  20
17        2   3   2   1   0   0   1   0  30
18        2   3   3   0   0   0   0   0   0
19        3   1   1   1   1   0   0   1  10
20        3   1   2   1   0   1   0   0  20
21        3   1   3   0   0   0   0   0   0
22        3   2   1   1   0   1   1   0  10
23        3   2   2   1   0   0   0   1  20
24        3   2   3   0   0   0   0   0   0
25        3   3   1   1   1   0   1   0  20
26        3   3   2   1   0   1   0   1  30
27        3   3   3   0   0   0   0   0   0
```

The design matrix contains categorical and/or continuous variables created by the `make.design.matrix()` function (i.e., `x2`, `x3`, `y2`, `y3`, and `Z` in this example) as well as the following four variables:

- `BLOCK`: The serial number of the blocks.
- `QES`: The serial number of the questions for each value of `BLOCK`.
- `ALT`: The serial number of alternatives for each value of `QES`.
- `ASC`: Alternative-specific constant(s). If the DCE design is labeled, the serial number of the alternatives is automatically appended to the tail of `ASC` (e.g., `ASC1`, `ASC2`, and `ASC3`).

The structure of the design matrix is such that each row shows one alternative; e.g., the first row shows the first alternative (`ALT = 1`) in the first question (`QES = 1`) in the first block (`BLOCK = 1`); the second row shows the second alternative (`ALT = 2`) in the first question (`QES = 1`) in the first block (`BLOCK = 1`); while the last row shows the third alternative (`ALT = 3`) in the third question (`QES = 3`) in the third block (`BLOCK = 3`).

The design matrix corresponds to the systematic component of utility for alternative i $(= 1, 2, 3)$ below:

$$\begin{aligned} V_1 &= ASC + \beta_1 x2_1 + \beta_2 x3_1 + \beta_3 y2_1 + \beta_4 y3_1 + \beta_5 Z_1, \\ V_2 &= ASC + \beta_1 x2_2 + \beta_2 x3_2 + \beta_3 y2_2 + \beta_4 y3_2 + \beta_5 Z_2, \\ V_3 &= 0, \end{aligned} \tag{3.8}$$

where β_1 is the coefficient of dummy variable $x2$ equal to `1` if attribute X is "x2," and `0` otherwise; β_2 is the coefficient of dummy variable $x3$ equal to `1` if attribute X is "x3," and `0` otherwise; β_3 is the coefficient of dummy variable $y2$ equal to `1` if attribute Y is "y2," and `0` otherwise; β_4 is the coefficient of dummy variable $y3$ equal to `1` if attribute Y is "y3," and `0` otherwise; and β_5 is the coefficient of variable Z for the continuous attribute Z.

See Sections 3.4.2 and 3.4.3 for examples using a labeled DCE and BDCE, respectively.

3.3.5 Creating a dataset

The `make.dataset()` function is able to create a dataset suitable for use by `clogit()` or `glm()` by combining the dataset containing information about the responses to the DCE questions and that containing the design matrix related to these questions. A generic call to the function and an explanation of its arguments are given below:

```
make.dataset(respondent.dataset,
             design.matrix,
             choice.indicators,
             detail = FALSE)
```

- `respondent.dataset`: A data frame containing the respondents' answers to the DCE questions.
- `design.matrix`: A data frame containing a design matrix created by `make.design.matrix()`.
- `choice.indicators`: A vector containing response variables showing the alternative that was selected in each DCE question.
- `detail`: A logical variable that denotes, if `TRUE`, that the variables contained in `respondent.dataset` and those created in this function are stored in a dataset produced by this function, and if `FALSE` that these variables are not stored.

The respondent dataset, which is assigned to `respondent.dataset`, must be created by the user. The dataset, in which each row depicts a single respondent, must contain variable `ID`, corresponding to the respondent's identification number; variable `BLOCK`, corresponding to the serial number of blocks to which each respondent is assigned; and response variables, corresponding to the answers to each of the DCE questions. If necessary, covariates showing the respondent's individual characteristics such as age and gender are also included in the respondent dataset. Although the names of the response variables and covariates are discretionary, the names of the respondent's identification number variable and block variable must be `ID` and `BLOCK`, respectively. In `respondent.dataset`, all variables are automatically treated as covariates, except for `ID`, `BLOCK`, and the response variables assigned by `choice.indicators`.

Table 3.1 *An example of an artificial respondent dataset (N = 30)*

ID	BLOCK	q1	q2	q3	ID	BLOCK	q1	q2	q3
1	1	2	1	2	16	1	2	1	1
2	2	2	2	1	17	2	2	2	1
3	3	2	3	2	18	3	1	1	2
4	1	1	1	1	19	1	1	1	3
5	2	2	2	1	20	2	2	2	2
6	3	1	2	2	21	3	1	2	3
7	1	1	1	1	22	1	1	1	3
8	2	2	2	3	23	2	2	2	2
9	3	1	1	1	24	3	1	1	1
10	1	3	3	1	25	1	3	1	1
11	2	2	2	1	26	2	2	2	2
12	3	1	2	2	27	3	1	1	2
13	1	3	3	1	28	1	3	3	1
14	2	2	2	3	29	2	2	2	3
15	3	1	1	2	30	3	1	1	3

For example, Table 3.1 shows an artificial respondent dataset containing 30 respondents' answers to the DCE questions created in Section 3.3.2. According to the conditions regarding the design of choice sets, it is assumed that each respondent answers a questionnaire containing an arbitrarily assigned block (any one of three) of a DCE design, each respondent is asked three DCE questions, each of which has three alternatives, and the responses are stored in variables `q1`, `q2`, and `q3`.

The answers to the DCE questions can be interpreted as follows: the first row shows that respondent 1 (`ID = 1`) was asked to respond to a questionnaire containing the first block (`BLOCK = 1`) of the DCE design and selected the second (`2`), first (`1`), and second (`2`) alternatives for the first (`q1`), second (`q2`), and third (`q3`) DCE questions, respectively. The last row shows that respondent 30 (`ID = 30`) was asked to respond to a questionnaire containing the third block (`BLOCK = 3`) of the DCE design and selected the first (`1`), first (`1`), and third (`3`) alternatives for the first (`q1`), second (`q2`), and third (`q3`) DCE questions, respectively.

As mentioned above, the response variables show the serial number of the alternative selected by the respondent for each DCE question. The method for assigning the serial number of the alternatives must be the same as that for assigning `ALT` in the output from `make.design.matrix()`. In other words, each alternative must be assigned an integer value starting at 1.

It should be noted that the order of the questions in the respondent dataset must be the same as that of variable `QES` in the design matrix dataset that was assigned to `design.matrix`, if the order of the DCE questions presented to respondents has been randomized.

Example code and output

After creating a respondent dataset (`res`) corresponding to Table 3.1 in R, `make.dataset()` is executed to create the dataset for analysis by combining the respondent dataset (`res`) and design matrix (`dm.rd`) that was created in Section 3.3.4:

```
R> res <- data.frame(ID = c(1:30),
                     BLOCK = rep(c(1:3), times = 10),
                     q1 = c(2, 2, 2, 1, 2, 1, 1, 2, 1, 3,
                            2, 1, 3, 2, 1, 2, 2, 1, 1, 2,
                            1, 1, 2, 1, 3, 2, 1, 3, 2, 1),
                     q2 = c(1, 2, 3, 1, 2, 2, 1, 2, 1, 3,
                            2, 2, 3, 2, 1, 1, 2, 1, 1, 2,
                            2, 1, 2, 1, 1, 2, 1, 3, 2, 1),
                     q3 = c(2, 1, 2, 1, 1, 2, 1, 3, 1, 1,
                            1, 2, 1, 3, 2, 1, 1, 2, 3, 2,
                            3, 3, 2, 1, 1, 2, 2, 1, 3, 3))
R> ds <- make.dataset(respondent.dataset = res,
                      design.matrix = dm.rd,
                      choice.indicators = c("q1", "q2", "q3"))

R> ds[1:10, ]
...
R> ds[268:270, ]
...
```

Figure 3.7 shows part of dataset `ds` created by executing the code given above. In addition to some variables included in the respondent and design matrix datasets, this dataset also contains the following two variables: `STR`, which is assigned to the argument `strata` in function `clogit()` to identify each combination of respondent and question, and `RES`, which is a logical variable that when used in `clogit()` takes the value `TRUE` if the alternative is selected and `FALSE` if it is not, whereas in `glm()` it takes the value `TRUE` when the first alternative is selected and `FALSE` when the second alternative is selected.

Because each row shows a single alternative, one DCE question has three alternatives, and each respondent is requested to answer three DCE questions, a group of nine rows shows one respondent's dataset; e.g., the first to the ninth rows give respondent 1's (`ID = 1`) responses to three DCE questions. From the first three rows, we see that respondent 1 was requested to respond to the first block of the DCE design (`BLOCK = 1`), faced with three alternatives in the first question (`QES = 1`) with the first one characterized by "x1" of attribute X, "y1" of attribute Y, and "10" of attribute Z; the second characterized by "x2" of attribute X, "y2" of attribute Y, and "20" of attribute Z; and the third as the opt-out option taking a value of 0 in all independent variables. Respondent 1 selected the second of the three alternatives (`RES` in row 2 = `TRUE`, while the

Respondent 1 selected second alternative

Each level of each attribute for first alternative

Each level of each attribute for second alternative

Each level of each attribute for third alternative

Block 1

Question 1

Respondent 1

	ID	BLOCK	QES	ALT	RES	ASC	x2	x3	y2	y3	z	STR
1	1	1	1	1	FALSE	1	0	0	0	0	10	101
2	1	1	1	2	TRUE	1	1	0	1	0	20	101
3	1	1	1	3	FALSE	0	0	0	0	0	0	101
4	1	1	2	1	TRUE	1	0	0	0	1	20	102
5	1	1	2	2	FALSE	1	1	0	0	0	30	102
6	1	1	2	3	FALSE	0	0	0	0	0	0	102
7	1	1	3	1	FALSE	1	0	1	0	1	30	103
8	1	1	3	2	TRUE	1	0	0	0	0	10	103
9	1	1	3	3	FALSE	0	0	0	0	0	0	103
10	2	2	1	1	FALSE	1	0	0	1	0	30	201
...												
268	30	3	3	1	FALSE	1	1	0	1	0	20	3003
269	30	3	3	2	FALSE	1	0	1	0	1	30	3003
270	30	3	3	3	TRUE	0	0	0	0	0	0	3003

From question 1 to 3 for respondent 1

Question 3 for respondent 30

Figure 3.7 *Example of a dataset suitable for analysis by* `clogit()` *(in the case of an unlabeled DCE).*

values of `RES` in rows 1 and 3 = `FALSE`). Similarly, the last three rows (rows 268–270) show the response of respondent 30 (`ID = 30`) to the third question.

See Section 3.4.3 for an example of a dataset for a BDCE.

3.3.6 Conducting statistical analysis

The `clogit()` function is used in a CL model analysis of the responses to DCE questions, whereas the `glm()` function is used in a BL model analysis of responses to BDCE questions.

A typical call to function `clogit()` together with an explanation of its arguments for the purpose of applying a CL model to the DCE responses is given below (see the reference manual for the details of arguments):

```
clogit(formula,
       data)
```

- `formula`: A model formula to be estimated.
- `data`: A data frame containing the dataset created using `make.dataset()`.

When applying the function, `formula` is set according to the DCE survey. The typical structure of `formula` for `clogit()` with K independent variables (`X`) including an alternative-specific constant (`ASC`) is defined as:

```
RES ~ ASC + X1 + X2 + ... + XK-1 + strata(STR)
```

where `RES` to the left of ~ is a logical variable equal to `TRUE` if the alternative is selected and `FALSE` if it is not. To the right of ~ are the independent variables and `strata(STR)`, which denotes that variable `STR` is used to identify each combination of respondent and question.

A typical call to the `glm()` function and an explanation of its arguments for the purpose of applying a BL model to the BDCE responses are given below (see the reference manual for the details of arguments):

```
glm(formula,
    family = binomial(link = "logit"),
    data)
```

- `formula`: A model formula to be estimated.
- `family`: The error distribution and link function.
- `data`: A data frame containing the dataset created by `make.dataset()`.

The `glm()` function is one for generalized linear models. Therefore, the argument `family` has to be set according to the preferred model; for the BL model, the argument is set to `family = binomial(link = logit)`. The `formula` is also set; the basic `formula` with $K - 1$ independent variables (`X`) is given as:

```
RES ~ X1 + X2 + ... + XK-1
```

In the above, `formula` does not contain `ASC`. This is because the `glm()` function automatically inserts a constant term in the model assigned to `formula`. If

ASC, which is contained in the dataset generated by `make.dataset()`, is used instead of the constant term, it is necessary to add `-1` and `ASC` to `formula` as shown in the following code, where `-1` deletes the constant term:

```
RES ~ -1 + ASC + X1 + X2 + ... + XK-1
```

Example code and output

As an example, the code for applying `clogit()` to analyze dataset `ds` can be written as:

```
R> fm <- RES ~ ASC + x2 + x3 + y2 + y3 + Z + strata(STR)
R> cl <- clogit(fm, data = ds)
R> cl
Call:
clogit(fm, data = ds)


       coef exp(coef) se(coef)      z       p
ASC   1.357     3.883   0.6018  2.254 2.4e-02
x2   -0.247     0.781   0.6099 -0.404 6.9e-01
x3    1.061     2.888   0.4070  2.606 9.2e-03
y2    1.436     4.205   0.6096  2.356 1.8e-02
y3    2.774    16.018   0.6001  4.622 3.8e-06
Z    -0.126     0.882   0.0313 -4.030 5.6e-05

Likelihood ratio test=73.4  on 6 df, p=8.35e-14  n= 270,
 number of events= 90
```

Columns `coef`, `exp(coef)`, `se(coef)`, `z`, and `p` give the estimated coefficient, exponential function of the estimated coefficient, standard error of the estimated coefficient, z-value, and p-value under the null hypothesis that the estimated coefficient is equal to zero for the corresponding independent variable shown in each row. A positive (negative) value of the estimated coefficient of an independent variable means that an increase in the value of the independent variable in an alternative increases (decreases) the systematic component of the utility for the alternative (see Eq. (3.2)).

Although a list of various components related to an estimated model is returned, the output above shows only formatted estimates and some summary statistics because a generic function `print()` has been defined specifically for the output from `clogit()`. For example, the estimated coefficients and their variance–covariance matrix in the output of `clogit()` are stored in the `coefficients` and `var` objects, respectively:

```
R> cl$coefficients
    ASC      x2      x3      y2      y3       Z
 1.3566 -0.2466  1.0606  1.4362  2.7737 -0.1261
R> cl$var
```

```
           [,1]     [,2]       [,3]     [,4]       [,5]
[1,]  0.36212 -0.11365 -0.0508176  0.00411  0.0271947
[2,] -0.11365  0.37203  0.1012776 -0.21705 -0.1470522
[3,] -0.05082  0.10128  0.1656202 -0.03254 -0.0004458
[4,]  0.00411 -0.21705 -0.0325435  0.37166  0.2639245
[5,]  0.02719 -0.14705 -0.0004458  0.26392  0.3600694
[6,] -0.01218  0.00557 -0.0018907 -0.00812 -0.0114802
           [,6]
[1,] -0.0121840
[2,]  0.0055700
[3,] -0.0018907
[4,] -0.0081196
[5,] -0.0114802
[6,]  0.0009783
```

Although the `var` matrix has no row or column names, the order of the rows (columns) is the same as that in `coefficients`: i.e., in this example, rows (columns) 1, 2, 3, 4, 5, and 6 correspond with `ASC`, `x2`, `x3`, `y2`, `y3`, and `Z`, respectively. As an example, the element in row 2 and column 2 shows the variance in the coefficient of `x2`, while that in row 2 and column 3 shows the covariance between the coefficients of `x2` and `x3`.

See Section 3.4.3 for example code and the resulting output for function `glm()`.

3.3.7 Calculating goodness-of-fit measures

The `gofm()` function can be used to evaluate the estimated discrete choice models. A generic call to the function is given below:

```
gofm(output)
```

where `output` is an object containing the output from either `clogit()` or `glm()`.

Example code and output

As an example, the following code returns measures related to the estimated model `cl` in Section 3.3.6:

```
R> gofm(cl)
Rho-squared = 0.371
Adjusted rho-squared = 0.3103
Akaike information criterion (AIC) = 136.4
Bayesian information criterion (BIC) = 151.4
Number of coefficients = 6
Log likelihood at start = -98.88
Log likelihood at convergence = -62.2
```

Because the function returns an object of the S3 class "`gofm`" and a `print()` function for the class has been defined, formatted output is shown in the example above. This class contains a list of the following components:

- `RHO2`: ρ^2.
- `AdjRHO2`: ρ^2 adjusted by the number of estimated coefficients.
- `AIC`: Akaike information criterion.
- `BIC`: Bayesian information criterion.
- `K`: The number of estimated coefficients.
- `LL0`: The log-likelihood value at the start.
- `LLb`: The log-likelihood value at convergence.

See Chapter 1 for definitions of these measures.

3.3.8 *Calculating MWTPs*

The `mwtp()` function calculates the MWTPs for attribute levels and confidence intervals of the MWTPs according to Krinsky and Robb, as well as delta methods. A generic call to `mwtp()` and definitions of its arguments are given below:

```
mwtp(output,
     monetary.variables,
     nonmonetary.variables = NULL,
     nreplications = 10000,
     confidence.level = 0.95,
     method = "kr",
     seed = NULL)
```

- `output`: An object containing the output from `clogit()` or `glm()`.
- `monetary.variables`: A vector containing the monetary variables in `output` used to calculate the MWTPs.
- `nonmonetary.variables`: A vector or list of vectors containing the non-monetary variables in `output` used to calculate the MWTPs.
- `nreplications`: An integer variable containing the number of replications in the simulation methods.
- `confidence.level`: A value showing the confidence level (coefficient) of an empirical distribution of each MWTP. The default value is set as `0.95`, which indicates the lower and upper bounds of the 95% confidence interval.
- `method`: A character variable describing the method used to calculate the confidence intervals of MWTPs: `"kr"` denotes that Krinsky and Robb's method is used, while `"delta"` implies that the Delta method is used.
- `seed`: Seed for the random number generator.

The definition of the MWTP of a non-monetary variable provided by `mwtp()` is the same as Eq. (3.5).

Two points should be noted with regard to assigning monetary and non-monetary variables in the output. First, when `nonmonetary.variables` is not set by the user, variables in `output`, except for those assigned by `monetary.variables`, are treated as non-monetary variables and the MWTPs for these variables are calculated.

Second, in a model that assumes alternative-specific attribute variables (that is, a labeled type DCE design), variables in `output` are classified into monetary and non-monetary variables according to the alternatives. Therefore, `monetary.variables` is set as a vector, while `nonmonetary.variables` is set as a list of vectors (see Section 3.4.2).

Example code and output

For example, executing the code given below returns the MWTPs for `ASC`, `x2`, `x3`, `y2`, and `y3`, which were estimated in Section 3.3.6:

```
R> mwtp.cl <- mwtp(output = cl,
                   monetary.variables=c("Z"),
                   nreplications = 1000,
                   confidence.level = 0.95,
                   method = "kr",
                   seed = 123)
R> mwtp.cl
      MWTP   2.5%  97.5%
ASC  10.76   2.29  18.52
x2   -1.96 -11.28   9.27
x3    8.41   2.15  18.21
y2   11.39   1.23  21.33
y3   22.00  13.82  34.38

method = Krinsky and Robb
```

The results show that MWTPs for `ASC`, `x3`, `y2`, and `y3` are significantly positive, whereas that for `x2` does not differ significantly from zero.

Because the function returns an object of the S3 class "`mwtp`" and a `print()` function for the class has been defined, the output in the above is formatted for ease of interpretation. An object of the class "`mwtp`" actually contains a list of the following components:

- `mwtp.table`: A matrix containing the MWTPs for the non-monetary attribute variables and confidence intervals for each of the MWTPs.
- `mwtps`: A matrix containing empirical distributions of the MWTPs.
- `repb`: A matrix containing `nreplications` sets of replicated coefficients.
- `method`: A character variable describing the method used to calculate the confidence intervals of the MWTPs.

Note that components `mwtps` and `rep` are not included in the output if the delta method is used to calculate the confidence intervals.

3.3.9 *Testing the difference between two independent MWTP distributions*

The `mded()` function can be used to test the difference between two independent empirical MWTP distributions, which are outputs from `mwtp()` based on the complete combinatorial method. A generic call to the function and definitions of its arguments are given below:

```
mded(distr1,
     distr2,
     detail = FALSE,
     independent = TRUE)
```

- `distr1, distr2`: Vectors of the respective empirical distributions, with `distr1` greater than `distr2`.
- `detail`: A logical variable that denotes, if `TRUE`, that a vector of the difference between `distr1` and `distr2` is returned, and if `FALSE` that no such vector is returned.
- `independent`: A logical variable that denotes, if `TRUE`, that `distr1` and `distr2` are independent of each other, and if `FALSE` that they are not independent of each other.

Two points should be noted. First, to use this function, two vectors of empirical MWTP distributions are needed (e.g., `mwtps` in the output from `mwtp()`). Second, executing `mded()` can take quite a long time and requires a large amount of memory to calculate the difference between two distributions if `detail` is set to `TRUE` because the resulting difference is stored as a vector. For example, if `distr1` and `distr2` each contain 10,000 elements (observations), the vector of the difference between the two distributions contains 100,000,000 elements. If insufficient memory is available, R terminates execution of the function, displaying an error message related to memory limitations.

Example code and output

As an example, the following code tests the difference between the MWTPs for `x2` and `x3` as calculated in Section 3.3.8:[12]

```
R> MWTP_x2 <- mwtp.cl$mwtps[, "x2"]
R> MWTP_x3 <- mwtp.cl$mwtps[, "x3"]
R> mded.cl <- mded(distr1 = MWTP_x3,
                   distr2 = MWTP_x2)
R> mded.cl
Test:
H0  MWTP_x3 = MWTP_x2
H1  MWTP_x3 > MWTP_x2
significance level = 0.0526
```

[12] Although the two MWTPs are not in fact independent, we consider them to be independent only for this example.

```
Data:
distr1 = MWTP_x3
distr2 = MWTP_x2

Means:
         means    n
MWTP_x3   8.76 1000
MWTP_x2  -1.65 1000

Cases in the difference:
             n
true     52555
false   947445
total  1000000
```

The resultant significance level is 0.0526, which implies that the null hypothesis of the equality between the MWTPs for `x2` and `x3` is rejected at the 10% level.

Because the output from `mded()` is an object of the S3 class "`mded`" and a `print()` function for this class has been defined, formatted output is shown in the example above. An object of the class "`mded`" contains a list of the following components:

- `stat`: One-sided significance level of the difference between `distr1` and `distr2`.
- `means`: A vector of the mean values of `distr1` and `distr2`.
- `cases`: A vector consisting of two integer values giving the number of cases in which subtracting `distr2` from `distr1` is negative and vice versa.
- `distr1, distr2`: Vectors assigned to `distr1` and `distr2`, respectively.
- `distr.names`: A vector of the names of objects assigned to `distr1` and `distr2`.
- `diff`: A vector of the difference. If `detail = TRUE`, this vector is returned.

3.4 Example DCEs using R

This section provides three examples showing how the functions discussed above can be used in DCEs and BDCEs. In these examples, defining a situation in which respondents select an alternative from the presented choice set and/or the attributes and their levels may not be sufficient because the examples are for illustrative purposes only, and the responses to the DCE questions in the examples have been artificially generated using random numbers. When applying DCEs in actual empirical research, the choice situations and definitions of attributes and their levels should be clearly explained to the respondents.

Q1. Please select one of the two types of rice that you would like to purchase.

	Rice 1	Rice 2
Region of origin	Region B	Region C
Cultivation method	NoChem	Organic
Price	JPY 2,000	JPY 2,300

[ ] I would like to purchase rice 1.
[ ] I would like to purchase rice 2.
[ ] I would like to purchase none of these.

Figure 3.8 *A choice situation example using an unlabeled design.*

3.4.1 Unlabeled DCE example

The first example relates to the use of an unlabeled DCE design, created by the mix-and-match method, in a questionnaire survey. We assume that a total of 100 consumers living in region B are requested to select their most preferred option from two alternatives (i.e., packages of rice) as well as the opt-out option (Figure 3.8).

Rice has three significant attributes:

- The region-of-origin attribute, which has three levels: "Region A" denoting a general rice production region, "Region B" denoting a general production region, but one in which the respondents live (that is, region B's rice is equivalent to locally produced rice for the respondents), and "Region C" denoting a well-known rice production region.
- The cultivation method attribute, which has three levels: "Conventional" denoting the conventional cultivation method, "NoChem" denoting zero usage of the agrichemical cultivation method, and "Organic" denoting an organic cultivation method.
- The price attribute, which gives the price per 5 kg of rice according to the three levels: "JPY 1,700," "JPY 2,000," and "JPY 2,300."[13]

Although the rice alternatives presented to respondents differ from each other in terms of the combination of the three attribute levels, other features of rice such as the variety, year harvested, and appearance are assumed to be the same for all rice alternatives.

The objective of applying a DCE in this example is to measure the respondents' relative importance of each level of the region of origin and the cultivation methods. Furthermore, the effect of the respondents' gender on their valuations of the cultivation methods is also examined.

The `rotation.design()` function with its arguments assigned according to the conditions given above is executed as follows:

[13] As at May 2013, USD 1 is approximately equivalent to JPY 100.

```
R> d.rice <- rotation.design(
                attribute.names = list(
                  Region = c("RegA", "RegB", "RegC"),
                  Cultivation = c("Conv", "NoChem", "Organic"),
                  Price = c("1700", "2000", "2300")),
                nalternatives = 2,
                nblocks = 1,
                row.renames = FALSE,
                randomize = TRUE,   # mix-and-match method
                seed = 987)
The columns of the array have been used in order of appearance.
For designs with relatively few columns,
the properties can sometimes be substantially improved
using option columns with min3 or even min34.

R> d.rice
Choice sets:
alternative 1 in each choice set
  BLOCK QES ALT Region Cultivation Price
1     1   1   1   RegB      NoChem  2000
8     1   2   1   RegB     Organic  1700
9     1   3   1   RegA     Organic  2000
6     1   4   1   RegA        Conv  1700
3     1   5   1   RegC     Organic  2300
7     1   6   1   RegA      NoChem  2300
5     1   7   1   RegC        Conv  2000
4     1   8   1   RegB        Conv  2300
2     1   9   1   RegC      NoChem  1700

alternative 2 in each choice set
  BLOCK QES ALT Region Cultivation Price
1     1   1   2   RegC     Organic  2300
5     1   2   2   RegA      NoChem  2300
8     1   3   2   RegC        Conv  2000
4     1   4   2   RegC      NoChem  1700
3     1   5   2   RegA        Conv  1700
9     1   6   2   RegB        Conv  2300
2     1   7   2   RegA     Organic  2000
7     1   8   2   RegB     Organic  1700
6     1   9   2   RegB      NoChem  2000

Candidate design:
  A B C
1 2 2 2
2 3 2 1
```

```
3 3 3 3
4 2 1 3
5 3 1 2
6 1 1 1
7 1 2 3
8 2 3 1
9 1 3 2
class=design, type= oa

Design information:
number of blocks = 1
number of questions per block = 9
number of alternatives per choice set = 2
number of attributes per alternative = 3
```

The number of blocks in the DCE design generated above is one and the size of the design is nine; that is, each respondent has to respond to a total of nine DCE questions, implying that the sample size of the analysis based on their responses is 900 (= 9 DCE questions per respondent × 100 respondents).

After creating the DCE design, the `questionnaire()` function is executed as follows (note that some output is omitted for the sake of clarity):

```
R> questionnaire(choice.experiment.design = d.rice)
Block 1

Question 1
            alt.1    alt.2
Region      "RegB"   "RegC"
Cultivation "NoChem" "Organic"
Price       "2000"   "2300"

...

Question 9
            alt.1    alt.2
Region      "RegC"   "RegB"
Cultivation "NoChem" "NoChem"
Price       "1700"   "2000"
```

Using the DCE questions created above, a questionnaire survey is carried out. In an actual questionnaire survey, the user of the DCE would have to create a respondent dataset based on the completed questionnaires; however, for the purpose of this example, the dataset was created and included in dataset `rice` in the **support.CEs** package as follows:

```
R> data(rice)
R> head(rice)
```

```
  ID BLOCK q1 q2 q3 q4 q5 q6 q7 q8 q9 F
1 1      1  1  1  2  2  1  3  2  2  1 0
2 2      1  1  1  2  1  1  3  2  2  1 1
3 3      1  2  2  3  2  3  3  1  2  2 0
4 4      1  2  1  2  2  3  2  1  2  1 1
5 5      1  1  1  2  2  1  1  1  2  3 1
6 6      1  2  1  1  2  3  1  2  3  1 0
```

The `rice` dataset, in which each row depicts the information of one respondent, contains variable `ID`, which shows the identification number of the respondent, variable `BLOCK`, which shows the block number, the response variables from `q1` to `q9`, and the respondent's gender variable `F` equal to `1` if the respondent is female, and `0` otherwise.

Next, the design matrix is created according to the DCE design. Here it is noted that the `Region` and `Cultivation` attributes are treated as categorical attributes, while `Price` is treated as a continuous attribute. The following code creates the design matrix and stores it in the object `dm.rice`:

```
R> dm.rice <- make.design.matrix(
     choice.experiment.design = d.rice,
     optout = TRUE,     # include opt-out option
     categorical.attributes = c("Region", "Cultivation"),
     continuous.attributes = c("Price"),
     unlabeled = TRUE) # unlabeled design
R> head(dm.rice)
  BLOCK QES ALT ASC RegB RegC NoChem Organic Price
1     1   1   1   1    1    0      1       0  2000
2     1   1   2   1    0    1      0       1  2300
3     1   1   3   0    0    0      0       0     0
4     1   2   1   1    1    0      0       1  1700
5     1   2   2   1    0    0      1       0  2300
6     1   2   3   0    0    0      0       0     0
```

The design matrix dataset contains variables `BLOCK`, `QES`, `ALT`, `ASC`, and other attribute variables, such as `RegB`, `RegC`, `NoChem`, `Organic`, and `Price`. Because there are three alternatives in each question in this example (see Figure 3.8), each group of three rows shows one choice set: the first to the third rows correspond to the first to third alternatives (`ALT` takes `1`, `2`, and `3` in the first, second, and third rows, respectively) of the first question (`QES = 1`) included in the first block of the DCE design (`BLOCK = 1`).

With the help of function `make.dataset()`, two datasets—`rice` and `dm.rice`—are combined to form dataset `ds.rice` for use by the `clogit()` function.

```
R> ds.rice <- make.dataset(
                 respondent.dataset = rice,
                 choice.indicators =
                   c("q1", "q2", "q3", "q4", "q5",
```

```
                         "q6", "q7", "q8", "q9"),
                    design.matrix = dm.rice)
R> head(ds.rice)
  ID F BLOCK QES ALT   RES ASC RegB RegC NoChem Organic
1  1 0     1   1   1  TRUE   1    1    0      1       0
2  1 0     1   1   2 FALSE   1    0    1      0       1
3  1 0     1   1   3 FALSE   0    0    0      0       0
4  1 0     1   2   1  TRUE   1    1    0      0       1
5  1 0     1   2   2 FALSE   1    0    0      1       0
6  1 0     1   2   3 FALSE   0    0    0      0       0
  Price STR
1  2000 101
2  2300 101
3     0 101
4  1700 102
5  2300 102
6     0 102
```

Dataset `ds.rice`, in which each group of three rows shows a choice set for a respondent, contains the variables included in `rice` and `dm.rice`, as well as the response variable (`RES`) and stratification variable (`STR`). For example, the first three rows correspond to the three alternatives in the first question (`QES = 1`) for respondent 1 (`ID = 1`); the values of `RES` in these three rows indicate that the respondent selected the first alternative in the first question (the value of `RES` in row 1 is `TRUE`, while the values in rows 2 and 3 are `FALSE`).

Before applying function `clogit()` to the dataset, a systematic component of the utility in the example is identified. Considering the effect of the respondents' gender on their valuation of non-usage of the agrichemical cultivation method or organic cultivation method, the systematic component of the utility of respondent n $(= 1, 2, ..., 100)$ in choosing rice alternative i $(= 1, 2)$ is expressed as follows (the systematic component of utility for the opt-out option is normalized to zero):

$$\begin{aligned} V_{in} &= ASC + \beta_1 RegB_{in} + \beta_2 RegC_{in} + \\ &\quad \beta_3 NoChem_{in} + \beta_4 Organic_{in} + \\ &\quad \beta_5 NoChem_{in} F_n + \beta_6 Organic_{in} F_n + \\ &\quad \beta_7 Price_{in}, \end{aligned} \tag{3.9}$$

where ASC refers to an alternative-specific constant; β_1 is the coefficient of variable $RegB$ equal to `1` if the region-of-origin attribute is "Region B," and `0` otherwise; β_2 is the coefficient of variable $RegC$ equal to `1` if the region-of-origin attribute is "Region C," and `0` otherwise; β_3 is the coefficient of variable $NoChem$ equal to `1` if the cultivation method is "NoChem," and `0` otherwise; β_4 is the coefficient of variable $Organic$ equal to `1` if the cultivation method is "Organic," and `0` otherwise; β_5 and β_6 are the coefficients of interaction between variable *NoChem* or *Organic*, respectively, and variable F equal to 1 if respondent n is female, and 0 otherwise; and β_7 is the coefficient of variable *Price* denoting the price per 5 kg of rice.

The `clogit()` function based on Eq. (3.9) and applied to dataset `ds.rice` is executed as follows:

```
R> fm.rice <- RES ~ ASC + RegB + RegC +
                    NoChem + Organic +
                    NoChem:F + Organic:F +
                    Price +
                    strata(STR)
R> out.rice <- clogit(fm.rice, data = ds.rice)
R> out.rice
Call:
clogit(fm.rice, data = ds.rice)


                coef exp(coef) se(coef)      z       p
ASC          4.44306    85.035 0.482971  9.199 0.0e+00
RegB         0.46909     1.599 0.137283  3.417 6.3e-04
RegC         0.96775     2.632 0.107571  8.996 0.0e+00
NoChem       0.75163     2.120 0.176572  4.257 2.1e-05
Organic      1.16466     3.205 0.141132  8.252 1.1e-16
Price       -0.00227     0.998 0.000233 -9.732 0.0e+00
NoChem:F    -0.19422     0.823 0.233636 -0.831 4.1e-01
Organic:F   -0.12266     0.885 0.183405 -0.669 5.0e-01

Likelihood ratio test=436  on 8 df, p=0  n= 2700, number of
events= 900
```

The goodness-of-fit measures calculated by function `gofm()` are given as follows:

```
R> gofm(out.rice)
Rho-squared = 0.2203
Adjusted rho-squared = 0.2122
Akaike information criterion (AIC) = 1558
Bayesian information criterion (BIC) = 1596
Number of coefficients = 8
Log likelihood at start = -988.8
Log likelihood at convergence = -770.9
```

According to the results of function `clogit()`, variables `RegB` and `RegC` have significantly positive coefficients, indicating that consumers value regions B and C higher than region A. All coefficients of variables `NoChem` and `Organic` are significantly positive, indicating that the values of non-usage of the agrichemical and organic cultivation methods is higher than those of the conventional cultivation method. However, all coefficients of variables `NoChem:F` and

Organic:F do not differ significantly from zero. The coefficient of variable Price is significantly negative, indicating consumers' preferences for cheaper rice. The value of the likelihood ratio test for the null hypothesis that all coefficients are zero is 436, which means that the null hypothesis is rejected. The adjusted ρ^2 value is over 0.2, meaning that the result is a good fit.

Now, let us examine whether the coefficient of RegB is statistically different from that of RegC. Here, the log-likelihood ratio test is used. The unrestricted model has been estimated in the above (i.e., out.rice). The restricted model in which the coefficient of RegB is equal to that of RegC, is estimated using function update() as follows:

```
R> out.rice.r <- update(out.rice,
                          .~. - RegB - RegC + I(RegB + RegC))
```

This code shows that the new model formula consists of the same variables as those in the model formula in out.rice (i.e., fm.rice) excluding the two variables RegB and RegB and with the addition of a new variable I(RegB + RegC) (i.e., the sum of RegB and RegC), which is a dummy variable equal to 1 if the rice is produced in region B or region C, and 0 otherwise. In the output from the clogit() function, the log-likelihood value at the start and that at convergence are stored in the object loglik. Thus, the log-likelihood ratio test value for our case is calculated as:

```
R> -2 * (out.rice.r$loglik[2] - out.rice$loglik[2])
[1] 16.13
```

The null hypothesis is rejected at the 1% significance level because the value of the statistic is greater than $\chi^2_{1,0.01} = 6.634$. This result implies that consumers in region B prefer rice that is produced in the well-known rice producing region over that which is produced locally.

Finally, for the first example, the MWTP for each non-monetary variable is calculated using the Krinsky and Robb method, which is coded using the mwtp() function as follows:

```
R> mwtp(output = out.rice,
        monetary.variables = c("Price"),
        nonmonetary.variables =
          c("RegB", "RegC", "NoChem", "Organic",
            "NoChem:F", "Organic:F"),
        confidence.level = 0.95,
        method = "kr",
        seed = 987)
              MWTP   2.5%  97.5%
RegB         207.0   83.8  348.4
RegC         427.1  321.7  560.7
NoChem       331.7  179.7  511.2
Organic      514.0  378.0  684.0
NoChem:F     -85.7 -290.4  118.7
```

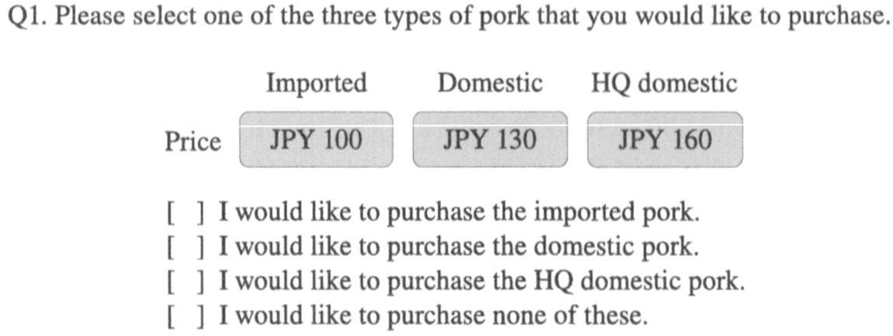

Figure 3.9 *Example of a choice situation using a labeled design.*

```
Organic:F  -54.1 -217.4  102.9

method = Krinsky and Robb
```

The output shows that compared with region A, the MWTPs for regions B and C are JPY 207.0 and JPY 427.1 per 5 kg of rice, and compared with the conventional cultivation method, the MWTPs for the non-usage of agrichemical and organic cultivation methods are JPY 331.7 and JPY 514.0 per 5 kg of rice. Although the MWTPs for `NoChem:F` and `Organic:F` are not significant, those for other variables are significantly positive at the 0.05 significance level.

3.4.2 Labeled design example

The second example relates to the use of a labeled DCE design, created by the L^{MA} design method, in a questionnaire survey. We assume that a total of 200 consumers are requested to select their most preferred option from three types of pork and the "none of these" option (Figure 3.9).

The pork alternative is expressed by two attributes:

- The country-of-origin attribute, which is treated as an alternative-specific attribute: the first, second, and third alternatives in a choice set are always "Imported," "Domestic," and "HQ (high quality) domestic," respectively.
- The price per 100 g attribute, which has four levels: "JPY 100," "JPY 130," "JPY 160," and "JPY 190."

Because the country-of-origin attribute is the alternative-specific attribute, the argument `attribute.names` in the `Lma.design()` function is set using only the price attribute.

According to the conditions mentioned above and the fact that the DCE design is assumed to be divided into 4 blocks, the `Lma.design()` function is executed as follows (some of the output is omitted for the sake of clarity):

```
R> d.pork <- Lma.design(
              attribute.names = list(
                Price = c("100", "130", "160", "190")),
              nalternatives = 3,
```

```
              nblocks = 4,
              row.renames = FALSE,
              seed = 987)
The columns of the array have been used in order of appearance.
For designs with relatively few columns,
the properties can sometimes be substantially improved
using option columns with min3 or even min34.

R> d.pork
Choice sets:
alternative 1 in each choice set
   BLOCK QES ALT Price
13     1   1   1   100
1      1   2   1   190
5      1   3   1   130
2      1   4   1   160
11     2   1   1   130
4      2   2   1   190
3      2   3   1   100
9      2   4   1   160
6      3   1   1   160
12     3   2   1   130
16     3   3   1   100
10     3   4   1   190
15     4   1   1   160
14     4   2   1   130
8      4   3   1   190
7      4   4   1   100

...

Design information:
number of blocks = 4
number of questions per block = 4
number of alternatives per choice set = 3
number of attributes per alternative = 1
```

The resulting DCE design shows that each respondent is requested to respond to four DCE questions.

DCE questions based on the DCE design are created using the `questionnaire()` function as follows (once again, some output is omitted):

```
R> questionnaire(choice.experiment.design = d.pork)
Block 1

Question 1
```

```
      alt.1 alt.2 alt.3
Price "100" "100" "100"

Question 2
      alt.1 alt.2 alt.3
Price "190" "160" "130"

...

Block 4

...

Question 4
      alt.1 alt.2 alt.3
Price "100" "130" "160"
```

Similar to the case in the first example, the responses to these DCE questions have already been created and stored in the **support.CEs** package as the dataset `pork`:

```
R> data(pork)
R> head(pork)
  ID BLOCK q1 q2 q3 q4
1  1     1  2  3  1  4
2  2     2  2  2  1  3
3  3     3  2  2  1  3
4  4     4  2  3  4  4
5  5     1  3  3  3  4
6  6     2  3  2  1  3
```

The dataset contains the four response variables (`q1`, `q2`, `q3`, and `q4`) corresponding to the four DCE questions, as well as the respondent's identification number variable (`ID`) and the block number variable (`BLOCK`).

Next, the design matrix is created according to the DCE design and stored in object `dm.pork` as shown below:

```
R> dm.pork <- make.design.matrix(
                choice.experiment.design = d.pork,
                optout = TRUE,      # include opt-out option
                continuous.attributes = c("Price"),
                unlabeled = FALSE) # labeled design
```

The two datasets, `pork` and `dm.pork`, are combined into one, `ds.pork`, using function `make.dataset()`, while at the same time assigning the vector of the four response variables to argument `choice.indicators`:

```
R> ds.pork <- make.dataset(
                respondent.dataset = pork,
```

```
                  choice.indicators =
                    c("q1", "q2", "q3", "q4"),
                  design.matrix = dm.pork)
```

In the second example, the systematic component of the utility of respondent n for alternative i (= 1 (Imported), 2 (Domestic), 3 (HQ domestic)) is modeled as shown below. (The systematic component of the utility for the "none of these" option is normalized to zero.)

$$V_{in} = ASC_i + \beta_i Price_{in}. \tag{3.10}$$

It should be noted that ASC and the coefficient of $Price$ are alternative-specific; in other words, these have a subscript i and can vary in value among alternatives.

After estimating the CL model based on Eq. (3.10) with dataset `ds.pork` by executing the `clogit()` function, the goodness-of-fit measures of the estimated model are calculated using the `gofm()` function:

```
R> fm.pork <- RES ~ ASC1 + Price1 + # variables for Imported
                    ASC2 + Price2 + # variables for Domestic
                    ASC3 + Price3 + # variables for HQ domestic
                    strata(STR)     # stratification variable
R> out.pork <- clogit(fm.pork, data = ds.pork)
R> out.pork
Call:
clogit(fm.pork, data = ds.pork)


          coef exp(coef) se(coef)      z       p
ASC1    4.1089    60.880  0.52586   7.81 5.6e-15
Price1 -0.0332     0.967  0.00415  -7.99 1.4e-15
ASC2    5.7114   302.309  0.42280  13.51 0.0e+00
Price2 -0.0370     0.964  0.00305 -12.13 0.0e+00
ASC3    4.5976    99.250  0.38405  11.97 0.0e+00
Price3 -0.0269     0.973  0.00262 -10.27 0.0e+00

Likelihood ratio test=526  on 6 df, p=0  n= 3200, number of
events= 800
R> gofm(out.pork)
Rho-squared = 0.2373
Adjusted rho-squared = 0.2319
Akaike information criterion (AIC) = 1704
Bayesian information criterion (BIC) = 1732
Number of coefficients = 6
Log likelihood at start = -1109
Log likelihood at convergence = -845.8
```

Each coefficient is significantly different from 0 at a significance level of 0.01 or less, and the resultant model is a good fit. Although cheaper pork is preferred because the coefficients of price for the three types of pork are negative, it is difficult to know which type of pork is preferred based on the estimated coefficients.

Accordingly, let us calculate the MWTP values for the different ASC values. In this example, the MWTP for ASC_i is the consumers' valuation of type i pork compared with the "none of these" option whose utility is zero by our assumption. The code required to calculate these MWTPs using the `mwtp()` function, and at the same time to assign the vector containing the three price variables to `monetary.variables` and the list of three components containing `ASC1`, `ASC2`, and `ASC3` to `nonmonetary.variables`, is given below.

```
R> mwtp.pork <- mwtp(
                  output = out.pork,
                  monetary.variables =
                    c("Price1", "Price2", "Price3"),
                  nonmonetary.variables = list(
                    c("ASC1"), c("ASC2"), c("ASC3")),
                  confidence.level = 0.95,
                  method = "kr",
                  seed = 987)
R> mwtp.pork
     MWTP 2.5% 97.5%
ASC1  124  116   132
ASC2  154  148   162
ASC3  171  162   182

method = Krinsky and Robb
```

The results show that the most preferred pork is the HQ domestic type (`ASC3`), the second most preferred pork is the domestic type (`ASC2`), and the least preferred is the imported type (`ASC1`). Figure 3.10 shows the empirical distribution of each MWTP for `ASC1`, `ASC2`, and `ASC3` stored in matrix `mwtps` included in `mwtp.pork`, which is drawn using the `hist()` function as follows:

```
R> hist(mwtp.pork$mwtps[, "ASC1"], main = "", xlab = "ASC1")
R> hist(mwtp.pork$mwtps[, "ASC2"], main = "", xlab = "ASC2")
R> hist(mwtp.pork$mwtps[, "ASC3"], main = "", xlab = "ASC3")
```

In the figure, the main title in each histogram is omitted by specifying `main = ""`, while each variable name is used as the label of the x-axis by setting `xlab =` to `"ASC1"`, `"ASC2"`, and `"ASC3"`, respectively.

Next, we conduct a likelihood ratio test for equality of the price coefficients of the three types of pork. In this case, the restricted model is a generic price attribute model, which means that the price variable in the three alternatives—`Price1`, `Price2`, and `Price3`—has a common coefficient. The unrestricted model, on the other hand, is an alternative-specific price

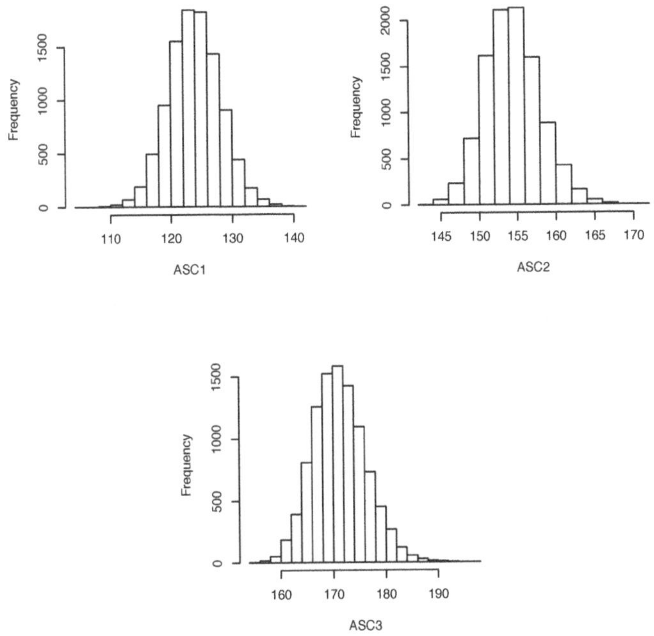

Figure 3.10 *MWTP histograms for* ***ASC1****,* ***ASC2****, and* ***ASC3****.*

attribute model, which means that the model allows a different coefficient of the price variable in the three alternatives. To estimate the generic price attribute model in this case, a generic `Price` variable is created in the dataset and then the model is estimated as:

```
R> ds.pork$Price <- ds.pork$Price1 + ds.pork$Price2 +
                    ds.pork$Price3
R> ds.pork[1:4, c("Price1", "Price2", "Price3", "Price")]
  Price1 Price2 Price3 Price
1    100      0      0   100
2      0    100      0   100
3      0      0    100   100
4      0      0      0     0
R> fm.porkg <- RES ~ ASC1 + ASC2 + ASC3 + # ASC for each pork
                     Price +      # generic price variable
                     strata(STR) # stratification variable
R> out.porkg <- clogit(fm.porkg, data = ds.pork)
R> out.porkg
Call:
clogit(fm.porkg, data = ds.pork)
```

```
          coef exp(coef) se(coef)       z p
ASC1    3.9416    51.500  0.26634   14.8 0
ASC2    5.0193   151.305  0.27760   18.1 0
ASC3    5.2768   195.734  0.28219   18.7 0
Price -0.0318      0.969  0.00184  -17.3 0

Likelihood ratio test=519  on 4 df, p=0  n= 3200, number of
events= 800
R> gofm(out.porkg)
Rho-squared = 0.2341
Adjusted rho-squared = 0.2305
Akaike information criterion (AIC) = 1707
Bayesian information criterion (BIC) = 1726
Number of coefficients = 4
Log likelihood at start = -1109
Log likelihood at convergence = -849.4
```

Log-likelihood values at the start and at convergence are stored in object `loglik` in the output from function `clogit()`: the first element in the object is the log-likelihood value at the start while the second is that at convergence.[14] The values in the alternative-specific price attribute model and those in the generic price attribute model are given as:

```
R> out.pork$loglik
[1] -1109.0  -845.8
R> out.porkg$loglik
[1] -1109.0  -849.4
```

Therefore, the likelihood ratio statistic is calculated as:

```
R> -2 * (out.porkg$loglik[2] - out.pork$loglik[2])
[1] 7.212
```

Because the value is greater than $\chi^2_{3,0.10} = 6.251$, the null hypothesis of equality of the price coefficients for the three alternatives is rejected at the 0.10 significance level.

Finally, for the second example, we predict the choice probability of each of the four alternatives by increasing the price of HQ domestic pork from JPY 100 to JPY 190 using the estimates from the alternative-specific price attribute model. Under the CL model, the probability of selecting alternative

[14] Log-likelihood values are also stored in the output from function `gofm()`. When extracting the value at convergence from the output, the output is assigned to any object (e.g., `gofm.pork <- gofm(out.portk)`) and then component `LLb` in the object is accessed (e.g., `gofm.pork$LLb`).

i from among the four alternatives is given as:

$$P_n(i) = \frac{\exp(V_{in})}{\sum_{j=1}^{4} \exp(V_{jn})}. \tag{3.11}$$

In addition to changing the HQ domestic price (`Price.hqd`), ranging from JPY 100 to JPY 190, it is also assumed that the price of imported pork (`Price.imp`) and that of domestic pork (`Price.dom`) is fixed at JPY 135 and JPY 150, respectively. These conditions can be expressed in R as shown below:

```
R> Price.imp <- c(135)
R> Price.dom <- c(150)
R> Price.hqd <- c(100:190)
```

Then, the systematic component of the utility for imported (`v.imp`), domestic (`v.dom`), and HQ domestic pork (`v.hqd`) are calculated using the estimated coefficients `coef`. Vector `coef` contains the estimated coefficients in the order of the variables as shown below:

```
R> out.pork$coef
    ASC1    Price1      ASC2    Price2      ASC3    Price3
 4.10891 -0.03317  5.71145 -0.03704  4.59764 -0.02689
```

Thus, the calculations are executed as:

```
R> v.imp <- out.pork$coef[1] + out.pork$coef[2] * Price.imp
R> v.dom <- out.pork$coef[3] + out.pork$coef[4] * Price.dom
R> v.hqd <- out.pork$coef[5] + out.pork$coef[6] * Price.hqd
```

According to Eq. (3.11), the choice probabilities of imported (`Prob.imp`), domestic (`Prob.dom`), and HQ domestic (`Prob.hqd`) pork, and that of the "none of these" option (`Prob.not`) are given as:

```
R> denominator.pork <- exp(v.imp) + exp(v.dom) + exp(v.hqd) + 1
R> Prob.imp <- exp(v.imp) / denominator.pork
R> Prob.dom <- exp(v.dom) / denominator.pork
R> Prob.hqd <- exp(v.hqd) / denominator.pork
R> Prob.not <- 1 - (Prob.imp + Prob.dom + Prob.hqd)
```

The results are shown in Figure 3.11, which is drawn using the following code:

```
R> plot(x = Price.hqd,                    # x-axis
        y = Prob.imp,                     # y-axis
        ylab = "Choice probability",      # label for y-axis
        ylim = c(0, 1),                   # limits for y-axis
        type = "l",                       # plot type is line
        lty = "solid")                    # line type is solid line
R> lines(Price.hqd, Prob.dom, # add dashed line of Prob.dom
         lty = "dashed")      #   on plot
R> lines(Price.hqd, Prob.hqd, # add dotted line of Prob.hqd
```

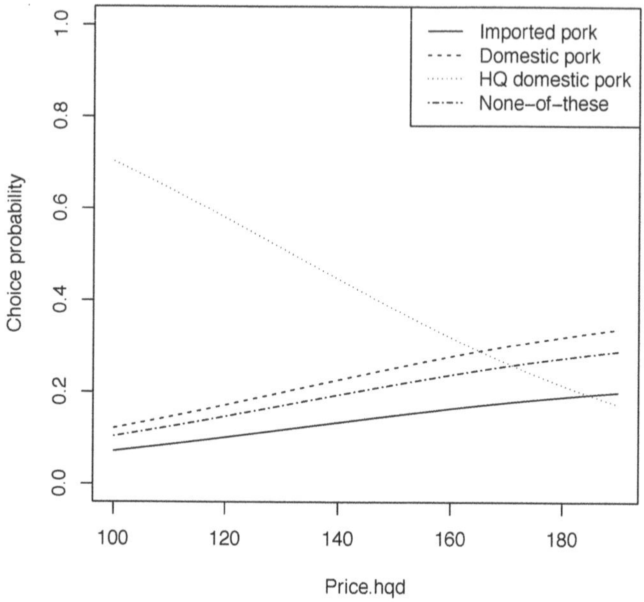

Figure 3.11 *Predicted probabilities of selecting the three types of pork and the "none of these" option.*

```
         lty = "dotted")       #  on plot
R> lines(Price.hqd, Prob.not, # add two-dashed line of Prob.not
         lty = "twodash")      #  on plot
R> legend("topright", # add legend at top right of plot
     legend = c("Imported pork",     # set names of each item
                "Domestic pork",
                "HQ domestic pork",
                "None-of-these"),
     lty = c("solid",  # set line types of each item
             "dashed", #  in the order corresponding to
             "dotted", #  that of the item names
             "twodash"))
```

Given that the prices of imported pork and domestic pork are JPY 135 and JPY 150, respectively, the choice probability of HQ domestic pork is higher than that of domestic pork if the price of HQ domestic pork is less than JPY 165.2, whereas the choice probability for HQ domestic pork surpasses that of imported pork if the HQ domestic price ranges between JPY 100 and JPY 184.7.

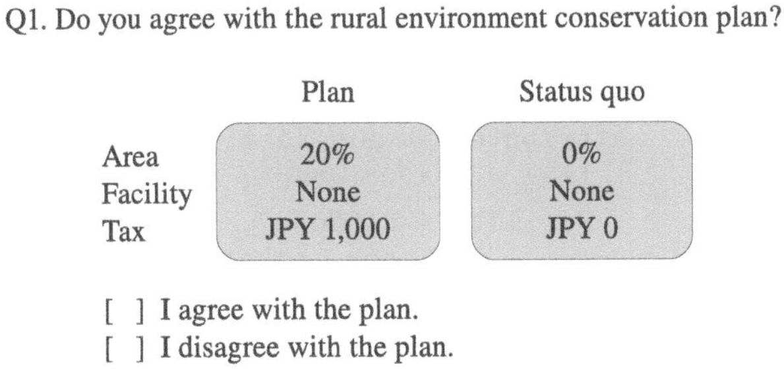

Figure 3.12 *Example of a choice situation using a BDCE.*

3.4.3 BDCE example

In the final example, a BDCE design is used in a questionnaire survey in which residents are required to evaluate a rural environment conservation plan composed of three attributes (Figure 3.12):

- An area attribute, which describes the targeted percentage of rural area in the region—"20%," "40%," "60%," and "80%."
- A facility attribute, which describes the type of facility constructed according to the plan—"None" denotes that no facility is constructed, "Agriculture (Agr)" denotes that a facility in which residents can learn features of agriculture in a rural area is constructed, "Environment (Env)" denotes that a facility in which residents can learn features of the environment in the rural area is constructed, and "Recreation (Rec)" denotes that a facility in which residents can enjoy recreation activities using the environment in the rural area is constructed.
- A tax attribute, which describes the additional amount of annual tax per household needed to carry out the plan—"JPY 1,000," "JPY 3,000," "JPY 5,000," and "JPY 7,000."

The area and tax attributes are treated as continuous attributes, while facility is treated as a categorical one.

In this example, two regions, the current rural environments of which are assumed to be the same, are each attempting to introduce the plan. Therefore, two different surveys need to be carried out; 200 residents in each region are requested to agree or disagree with the plan. If the respondents disagree with it, the status quo—with no targeted area, no facility, and no additional tax—is assumed to continue; the current situation is interpreted as the common base alternative. The main objective of applying a BDCE in this example is to measure the marginal willingness to pay for the attribute levels and compare the values for the two regions.

While ensuring that argument `nalternatives` is set to `1` and assuming that the BDCE design is divided into 4 blocks, the `Lma.design()` function constructs choice sets for this type of BDCE as follows (as before, some of the output is omitted):

```
R> d.rural <- Lma.design(
                  attribute.names = list(
                    Area = c("20", "40", "60", "80"),
                    Facility = c("None", "Agr", "Env", "Rec"),
                    Tax = c("1000", "3000", "5000", "7000")),
                  nalternatives = 1, # create design for a BDCE
                  nblocks = 4,
                  row.renames = FALSE,
                  seed = 987)
The columns of the array have been used in order of appearance.
For designs with relatively few columns,
the properties can sometimes be substantially improved
using option columns with min3 or even min34.

R> d.rural
Choice sets:
alternative 1 in each choice set
   BLOCK QES ALT Area Facility  Tax
13     1   1   1   20     None 1000
1      1   2   1   80      Env 3000
5      1   3   1   40      Rec 5000
2      1   4   1   60      Agr 7000
11     2   1   1   40      Agr 3000
4      2   2   1   80     None 5000
3      2   3   1   20      Env 7000
9      2   4   1   60      Rec 1000
6      3   1   1   60      Env 5000
12     3   2   1   40     None 7000
16     3   3   1   20      Rec 3000
10     3   4   1   80      Agr 1000
15     4   1   1   60     None 3000
14     4   2   1   40      Env 1000
8      4   3   1   80      Rec 7000
7      4   4   1   20      Agr 5000

Candidate design:
   A B C D
1  4 3 2 1
2  3 2 4 1
3  1 3 4 2
4  4 1 3 2
5  2 4 3 1
6  3 3 3 3
```

```
7  1 2 3 4
8  4 4 4 4
9  3 4 1 2
10 4 2 1 3
11 2 2 2 2
12 2 1 4 3
13 1 1 1 1
14 2 3 1 4
15 3 1 2 4
16 1 4 2 3
class=design, type= oa

Design information:
number of blocks = 4
number of questions per block = 4
number of alternatives per choice set = 1
number of attributes per alternative = 3
```

The resulting BDCE design shows that each respondent is presented with four choice situations from a total of 16 choice situations.

To create questions for this case using the `questionnaire()` function, argument `common` is set according to the features of the common base option (0%, None, and JPY 0). With this setting, the `questionnaire()` function returns questions with the common base alternative:

```
R> common.alt <- c(Area = "0", Facility = "None", Tax = "0")
R> questionnaire(choice.experiment.design = d.rural,
                 common = common.alt)
Block 1

Question 1
          alt.1  alt.2
Area      "20"   "0"
Facility "None" "None"
Tax      "1000" "0"

...

Block 4

...

Question 4
          alt.1  alt.2
Area      "20"   "0"
```

```
Facility "Agr"  "None"
Tax      "5000" "0"
```

The responses to the BDCE questions are stored in the dataset `rural` in the **support.CEs** package:

```
R> data(rural)
```

The dataset contains a variable `Region`, which denotes whether the respondent was sampled from region 1 (`Region = 1`) or region 2 (`Region = 2`). According to the value of `Region`, the dataset is divided into two subsets:

```
R> # extract rows with Region = 1
R> rural1 <- subset(rural, Region == 1)
R> # extract rows with Region = 2
R> rural2 <- subset(rural, Region == 2)
R> head(rural1)
  ID BLOCK q1 q2 q3 q4 Region
1  1     1  1  1  0  0      1
2  2     2  1  0  0  1      1
3  3     3  1  0  0  1      1
4  4     4  0  1  0  0      1
5  5     1  0  1  0  0      1
6  6     2  1  1  0  1      1
R> head(rural2)
     ID BLOCK q1 q2 q3 q4 Region
201 201     1  0  1  0  0      2
202 202     2  1  0  0  1      2
203 203     3  0  0  1  1      2
204 204     4  1  1  0  0      2
205 205     1  0  1  0  0      2
206 206     2  1  1  0  1      2
```

Similar to `questionnaire()`, the `make.design.matrix()` function with argument `common` set according to the common base alternative is executed as follows:

```
R> dm.rural <- make.design.matrix(
     choice.experiment.design = d.rural,
     optout = FALSE,        # do not include opt-out option
     categorical.attributes = c("Facility"),
     continuous.attributes = c("Area", "Tax"),
     unlabeled = TRUE,      # unlabeled design
     common = common.alt,   # include common alternative
     binary = TRUE)         # BDCEs
R> head(dm.rural)
  BLOCK QES ALT ASC Agr Env Rec Area  Tax
1     1   1   1   1   0   0   0   20 1000
2     1   2   1   1   0   1   0   80 3000
```

```
3       1  3  1  1  0  0  1    40 5000
4       1  4  1  1  1  0  0    60 7000
5       2  1  1  1  1  0  0    40 3000
6       2  2  1  1  0  0  0    80 5000
```

The structure of this design matrix is slightly different from that used in (multinomial) DCEs; that is, each row in the output above shows one question rather than one alternative. This is because the analysis of a BDCE using the `glm()` function needs independent variables that are calculated from the difference between the systematic component of utility for alternative i and that for alternative j. In this example, i is the proposed plan alternative and j is the common base option in which all attribute variables are equal to 0; thus V_j results in 0. Therefore, the values of the five attribute variables in this example denote levels in alternative i in the questions. In other words, `Agr` is equal to `1` if the plan constructs a facility used for learning agriculture in the region, and `0` otherwise; `Env` is equal to `1` if the plan constructs a facility used for learning the rural environment in the region, and `0` otherwise; `Rec` is equal to `1` if the plan constructs a recreation facility in the region, and `0` otherwise; and `Area` and `Tax` are continuous variables denoting the percentage of rural area targeted by the plan and amount of annual tax per household required to carry out the plan, respectively. Thus, the difference between V_i and V_j in this example can be expressed as:

$$V_i - V_j = ASC_i + \beta_1 Agr_i + \beta_2 Env_i + \beta_3 Rec_i + \beta_4 Area_i + \beta_5 Tax_i, \quad (3.12)$$

where each β is a coefficient corresponding to one of the variables. The independent variables—ASC, Agr, Env, Rec, $Area$, and Tax—are the same in the design matrix given above.

It is worth noting that the design matrix given above is also created by setting `optout = TRUE` and `common = NULL`:

```
R> dm.ruralop <- make.design.matrix(
     choice.experiment.design = d.rural,
     optout = TRUE, # include opt-out option
     categorical.attributes = c("Facility"),
     continuous.attributes = c("Area", "Tax"),
     unlabeled = TRUE, # unlabeled design
     common = NULL,     # do not include common alternative
     binary = TRUE)     # BDCEs
R> # test whether dm.rural is equal to dm.ruralop
R> identical(dm.rural, dm.ruralop)
[1] TRUE
```

This is because all attributes corresponding to the common base alternative in this example have a value of 0.

Datasets corresponding to regions 1 and 2 and suitable for use by the `glm()` function are created by combining the respondent dataset (`rural1` or rural2) and the design matrix (dm.rural) using the make.dataset() function:

```
R> ds.rural1 <- make.dataset(
     respondent.dataset = rural1,
     design.matrix = dm.rural,
     choice.indicators = c("q1", "q2", "q3", "q4"),
     detail = FALSE)
R> ds.rural2 <- make.dataset(
     respondent.dataset = rural2,
     design.matrix = dm.rural,
     choice.indicators = c("q1", "q2", "q3", "q4"),
     detail = FALSE)
R> head(ds.rural1)
  ID Region BLOCK QES ALT   RES ASC Agr Env Rec Area  Tax
1  1      1     1   1   1  TRUE   1   0   0   0   20 1000
2  1      1     1   2   1  TRUE   1   0   1   0   80 3000
3  1      1     1   3   1 FALSE   1   0   0   1   40 5000
4  1      1     1   4   1 FALSE   1   1   0   0   60 7000
5  2      1     2   1   1  TRUE   1   1   0   0   40 3000
6  2      1     2   2   1 FALSE   1   0   0   0   80 5000
  STR
1 101
2 102
3 103
4 104
5 201
6 202
```

Similar to the design matrix, the row structure of the dataset (where each row gives the data for one question per respondent) differs from that for DCEs (where each row gives the data for one alternative per question per respondent). However, the categories of variables are the same as those for DCEs: RES is a dependent variable, variables ASC to Tax are independent variables with respect to attributes/attribute levels, Region is an independent variable with respect to respondents' characteristics, and BLOCK, QES, and ALT provide information about which questions should be presented to each respondent. The STR variable is not used by glm().

Next, to estimate the BL model based on region 1's dataset and evaluate the estimated model, the glm() and gofm() functions are executed as follows:

```
R> fm.rural <- RES ~ Agr + Env + Rec + Area + Tax
R> out.rural1 <- glm(fm.rural,
                     family = binomial(link = "logit"),
                     data = ds.rural1)
R> summary(out.rural1)
```

```
Call:
glm(formula = fm.rural, family = binomial(link = "logit"),
data = ds.rural1)
Deviance Residuals:
   Min       1Q  Median      3Q     Max
-3.280  -0.441  -0.029   0.191   2.595

Coefficients:
             Estimate Std. Error z value Pr(>|z|)
(Intercept) -0.289892   0.327686   -0.88  0.37634
Agr          1.239121   0.407007    3.04  0.00233
Env          1.327223   0.402862    3.29  0.00099
Rec          1.392455   0.497268    2.80  0.00511
Area         0.113689   0.013177    8.63  < 2e-16
Tax         -0.001586   0.000183   -8.67  < 2e-16

(Dispersion parameter for binomial family taken to be 1)

    Null deviance: 1108.96  on 799  degrees of freedom
Residual deviance:  477.42  on 794  degrees of freedom
AIC: 489.4

Number of Fisher Scoring iterations: 8
R> gofm(out.rural1)
Rho-squared = 0.5695
Adjusted rho-squared = 0.5587
Akaike information criterion (AIC) = 489.4
Bayesian information criterion (BIC) = 517.5
Number of coefficients = 6
Log likelihood at start = -554.5
Log likelihood at convergence = -238.7
```

Functions glm() and gofm() for region 2 are also executed as follows:

```
R> out.rural2 <- glm(fm.rural,
                     family = binomial(link = "logit"),
                     data = ds.rural2)
R> summary(out.rural2)
Call:
glm(formula = fm.rural, family = binomial(link = "logit"),
data = ds.rural2)
Deviance Residuals:
   Min       1Q  Median      3Q     Max
-2.525  -0.342  -0.043   0.290   3.743

Coefficients:
```

```
             Estimate Std. Error z value Pr(>|z|)
(Intercept) -0.10530    0.33598   -0.31   0.7540
Agr          1.09105    0.37438    2.91   0.0036
Env          0.79057    0.34356    2.30   0.0214
Rec          0.32631    0.41762    0.78   0.4346
Area         0.09582    0.01016    9.43   <2e-16
Tax         -0.00137    0.00013  -10.57   <2e-16

(Dispersion parameter for binomial family taken to be 1)

    Null deviance: 1105.91  on 799  degrees of freedom
Residual deviance:  473.07  on 794  degrees of freedom
AIC: 485.1

Number of Fisher Scoring iterations: 7
R> gofm(out.rural2)
Rho-squared = 0.5734
Adjusted rho-squared = 0.5626
Akaike information criterion (AIC) = 485.1
Bayesian information criterion (BIC) = 513.2
Number of coefficients = 6
Log likelihood at start = -554.5
Log likelihood at convergence = -236.5
```

Variable `Area` in both regions has a significantly positive coefficient, indicating that residents prefer a higher ratio of targeted area. Variable `Tax` has a significantly negative coefficient, indicating that residents prefer a lower additional tax to implement the plan. Although variables `Agr`, `Env`, and `Rec` differ significantly from zero in region 1, only variables `Agr` and `Env` differ significantly from zero in region 2.

Next, we examine whether the MWTPs for non-monetary variables in region 1 are statistically different from those in region 2 using the `mded()` function.

First, the MWTPs in each region are calculated using the `mwtp()` function:

```
R> mwtp.rural1 <- mwtp(output = out.rural1,
                       monetary.variables = c("Tax"),
                       nonmonetary.variables =
                         c("Agr", "Env", "Rec", "Area"),
                       nreplications = 1000,
                       confidence.level = 0.95,
                       method = "kr",
                       seed = 987)
R> mwtp.rural1
```

```
       MWTP   2.5%  97.5%
Agr   781.3  306.2 1200.7
Env   836.8  392.6 1212.6
Rec   878.0  319.5 1285.1
Area   71.7   65.2   78.4

method = Krinsky and Robb
R> mwtp.rural2 <- mwtp(output = out.rural2,
                       monetary.variables = c("Tax"),
                       nonmonetary.variables =
                         c("Agr", "Env", "Rec", "Area"),
                       nreplications = 1000,
                       confidence.level = 0.95,
                       method = "kr",
                       seed = 987)
R> mwtp.rural2
       MWTP   2.5%  97.5%
Agr   794.9  304.7 1279.0
Env   576.0   76.5 1028.4
Rec   237.7 -400.2  730.8
Area   69.8   62.1   77.2

method = Krinsky and Robb
```

Then, the mded() function is used to test the difference between the MWTP in region 1 and that in region 2 based on the empirical distributions returned by the mwtp() function. The mwtps in the output mwtp.rural2 is a 1000 × 4 matrix in which the first, second, third, and fourth columns represent the empirical distributions of the MWTP for Agr, Env, Rec, and Area, respectively:

```
R> str(mwtp.rural1$mwtps)
 num [1:1000, 1:4] 1018 805 667 980 593 ...
 - attr(*, "dimnames")=List of 2
  ..$ : NULL
  ..$ : chr [1:4] "Agr" "Env" "Rec" "Area"
```

Here we consider the MWTPs for Rec and Area. Function mded() with arguments distr1 and distr2 representing the empirical distributions of the MWTP for Rec in regions 1 and 2, respectively, is executed as shown below:

```
R> MWTP_Rec1 <- mwtp.rural1$mwtps[, "Rec"]
R> MWTP_Rec2 <- mwtp.rural2$mwtps[, "Rec"]
R> mded.rec <- mded(distr1 = MWTP_Rec1,
                    distr2 = MWTP_Rec2,
                    detail = TRUE)
R> mded.rec
```

```
Test:
H0  MWTP_Rec1 = MWTP_Rec2
H1  MWTP_Rec1 > MWTP_Rec2
significance level = 0.0459

Data:
distr1 = MWTP_Rec1
distr2 = MWTP_Rec2

Means:
           means     n
MWTP_Rec1    856  1000
MWTP_Rec2    215  1000

Cases in the difference:
            n
true    45921
false  954079
total 1000000
```

Because the significance level is 0.0459, the null hypothesis that the MWTP for Rec in region 1 is equal to that in region 2 is rejected. See also the left-hand side of Figure 3.13, which is drawn as follows:

```
R> hist(mded.rec$diff, main = "(a) Rec", xlab = "Difference")
```

Similarly, the test for Area is executed as follows:

```
R> MWTP_Area1 <- mwtp.rural1$mwtps[, "Area"]
R> MWTP_Area2 <- mwtp.rural2$mwtps[, "Area"]
R> mded.area <- mded(distr1 = MWTP_Area1,
                     distr2 = MWTP_Area2,
                     detail = TRUE)
R> mded.area
Test:
H0  MWTP_Area1 = MWTP_Area2
H1  MWTP_Area1 > MWTP_Area2
significance level = 0.357

Data:
distr1 = MWTP_Area1
distr2 = MWTP_Area2

Means:
            means     n
MWTP_Area1   71.6  1000
MWTP_Area2   69.7  1000
```

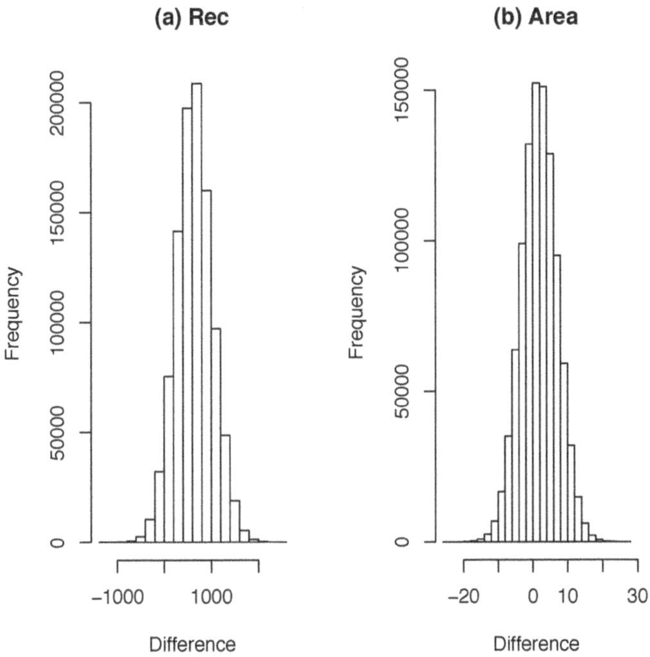

Figure 3.13 *Histograms showing the differences in the MWTP for* `Rec` *and for* `Area` *in the two regions.*

```
Cases in the difference:
            n
true   357291
false  642709
total 1000000
```

The result shows that the null hypothesis is not rejected. See also the right-hand side of Figure 3.13, which is drawn as follows:

```
R> hist(mded.area$diff, main = "(b) Area", xlab = "Difference")
```

Finally, for the last example, we calculate the economic values for the rural environment conservation plans in region 1. Consider the following three cases:

- A change from the status quo (the common base alternative) to plan 1 in which none of the facilities is constructed and 20% of the rural area is targeted.
- A change from the status quo to plan 2 in which a facility for learning the agriculture in the region is constructed and 60% of the rural area is targeted.
- A change from the situation in which the status quo and plan 1 alternatives

are available to a different situation in which the status quo and plan 2 alternatives are available.

The first and second cases correspond to the "state of the world model": the economic values are calculated using Eq. (3.7).

Because the values of `Agr`, `Env`, `Rec`, `Area`, and `Tax` are all set to `0` in the common base alternative, the value of the systematic component of the utility for the status quo is set to `0`:

```
R> v0 <- 0
```

Plan 1 is expressed as a vector based on the model formula used in the estimation:

```
R> Plan1 <- c(1, 0, 0, 0, 20, 0)
```

where the first element corresponds to the `(Intercept)` and the second to sixth elements correspond to variables `Agr`, `Env`, `Rec`, `Area`, and `Tax`, respectively. The systematic component of the utility for plan 1 is calculated as follows:

```
R> v1 <- sum(out.rural1$coef * Plan1)
```

According to the settings given above, the economic value of a change from the status quo to plan 1 is calculated as shown below:

```
R> (-1 / -out.rural1$coef["Tax"]) * (v0 - v1)
 Tax
1251
```

This means that the benefit associated with the change is JPY 1,251 per household per year.

Next, plan 2 represents the situation in which a facility for learning the agriculture in the region is constructed and 60% of the rural area is targeted; therefore, a vector corresponding to these conditions is created as follows:

```
R> Plan2 <- c(1, 1, 0, 0, 60, 0)
```

The systematic component of the utility for plan 2 and the economic value of a change from the status quo to plan 2 are calculated as shown below:

```
R> v2 <- sum(out.rural1$coef * Plan2)
R> (-1  / -out.rural1$coef["Tax"]) * (v0 - v2)
 Tax
4900
```

The confidence intervals for the economic values of the changes mentioned above can be calculated using the method of Krinsky and Robb (1986, 1990), which consists of the following three steps (see Chapter 1):

Step 1 A total of $R = 1000$ sets of coefficients of the estimated model are randomly drawn from a multivariate normal distribution using function `mvrnorm()` in the **MASS** package. The code is as follows:

```
R> set.seed(123)
R> COEF <- mvrnorm(out.rural1$coef,  # coefficients
                   vcov(out.rural1), # variance-covariance
                   n = 1000)         # number of replications
```

where the `vcov()` function calculates a variance–covariance matrix of the estimated model.

Step 2 An empirical distribution of each of the measures is computed from the R sets of randomly sampled coefficients. Here, economic values in the first (`C1`) and second (`C2`) cases are calculated as:

```
R> V0 <- rep(0, 1000)
R> V1 <- COEF %*% Plan1 # %*% is matrix multiplication
R> C1 <- (-1 / -COEF[, "Tax"]) * (V0 - V1)
R> V2 <- COEF %*% Plan2
R> C2 <- (-1  / -COEF[, "Tax"]) * (V0 - V2)
```

and their histograms are obtained using the code given below (Figure 3.14):

```
R> hist(C1, main = "Case 1", xlab = "Compensating variation")
R> hist(C2, main = "Case 2", xlab = "Compensating variation")
```

Step 3 A confidence interval for each of the values is constructed based on the respective empirical distribution. Here, 95% confidence intervals are calculated using the `quantile()` function:

```
R> quantile(C1, c(0.025, 0.975))
  2.5%  97.5%
 945.2 1597.2
R> quantile(C2, c(0.025, 0.975))
 2.5% 97.5%
 4546  5274
```

These results show that the economic values are significantly different from 0 at the 5% significance level.

We calculate the economic value of the third case based on Eq. (3.6); that is, a change from the situation in which the status quo and plan 1 alternatives are available to a different situation in which the status quo and plan 2 alternatives are available. Because all values of the systematic components of the utility for plan 1, plan 2, and the status quo have been calculated above, the economic value of the third case is easily calculated using the following code:

```
R> (-1 / -out.rural1$coef["Tax"]) *
     (log(exp(v0) + exp(v1)) - log(exp(v0) + exp(v2)))
 Tax
3568
```

Furthermore, the code given below generates the empirical distribution of the economic value in the third case, draws the distribution (Figure 3.15), and calculates the 95% confidence intervals based on the method of Krinsky and Robb (1986, 1990):

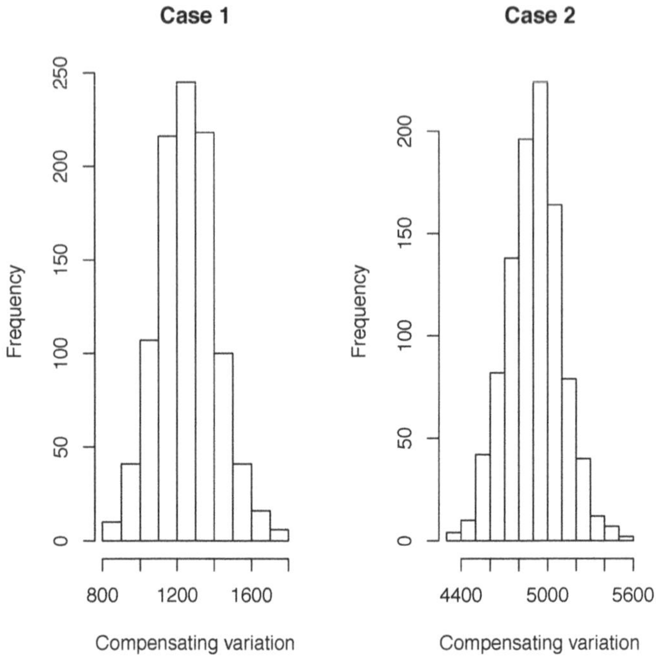

Figure 3.14 *Histograms of the simulated economic values in the first and second cases.*

```
R> C3 <- (-1 / -COEF[ ,"Tax"]) *
           (log(exp(V0) + exp(V1)) - log(exp(V0) + exp(V2)))
R> hist(C3, main = "", xlab = "Compensating variation")
R> quantile(C3, c(0.025, 0.975))
 2.5% 97.5%
 3047  4034
```

These results show that the economic value of the third case is JPY 3,568 per household per year, which is significantly different from 0 at the 0.05 significance level.

3.5 Concluding remarks

In this chapter we showed how various R functions can be used to implement basic DCEs. Although a certain amount of effort is necessary to conduct a survey and analysis to actualize the appropriate applications for DCEs, it is clear that the functions introduced in this chapter contribute significantly to

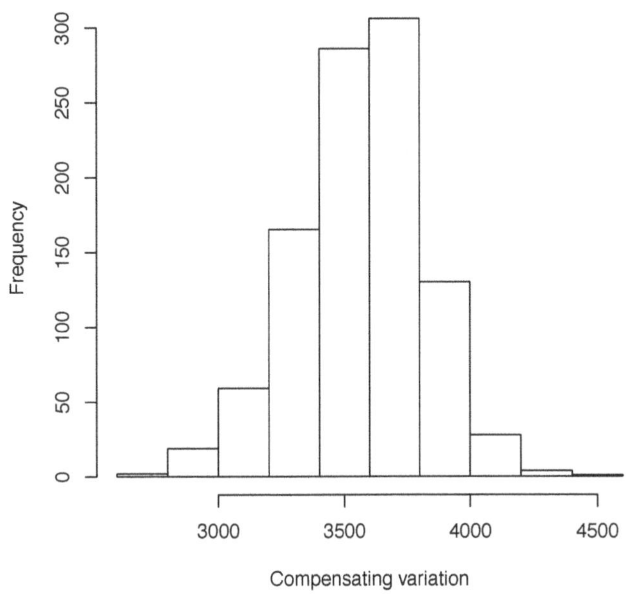

Figure 3.15 *Histograms of simulated economic values in the third case.*

the overall process, particularly for DCE beginners. Readers who are interested in applying advanced discrete choice models to DCEs should refer to Appendix A, in which various R packages related to discrete choice models are introduced.

Chapter 4

Best–Worst Scaling

Abstract

This chapter explains how best–worst scaling (BWS) is implemented in R. Although there are three types of BWS, namely, object case BWS, profile case BWS, and multiprofile case BWS, this chapter only addresses the first type. After discussing the basic structure and features of BWS, Section 4.2 explains two methods for constructing choice sets for BWS, including an approach using a two-level orthogonal main effect design and one using a balanced incomplete block design, as well as two methods for analyzing responses to BWS questions, that is, an approach counting the number of times each item is selected as the best and the worst, and one applying a conditional logit model. Section 4.3 describes how to implement these methods in R. Section 4.4 provides two examples of performing BWS in R, namely, a measure of the relative importance of the multifunctionality of agriculture and a consumers' evaluation of fruits.

4.1 Introduction

Best–worst scaling (BWS), or maximum difference scaling, is a survey method for eliciting individuals' relative importance of items (Finn and Louviere, 1992). BWS was developed to expand the capability of discrete choice experiments (DCEs) (Flynn et al., 2010). Although BWS is categorized into three types based on the format of the choice sets (Flynn et al., 2010; Marley, 2010), this chapter only addresses one of these in detail, namely, object case BWS. The other two types, profile case BWS and multiprofile case BWS, are briefly outlined in the appendix to this chapter. (Hereafter, BWS refers to object case BWS.)

BWS requires respondents to select two items from a choice set containing three or more items. The choices reflect the extremes of a certain standard, such as "best" and "worst," or "most important" and "least important." The BWS users can decide what standards should be used. Items can also be selected relatively flexibly. In DCE studies, goods/services are treated as alternatives and the characteristics of the goods/services are set as attributes of the alternatives (see Chapter 3). Although the characteristics of the alternatives are set as items in BWS, the respondents' views, experiences, and/or benefits related to the goods/services are also frequently set as items. For example,

Cohen (2009) used 13 items as possible factors affecting consumers' choice of wine. The items included respondents' experiences such as "I read about it in a guide" or "Tasted the wine previously," as well as characteristics of the wine such as grape variety and origin. Cohen and Neira (2004) applied BWS to measure the relative importance consumers placed on 13 benefits of drinking coffee, such as "It makes me feel relaxed" and "It helps me keep warm." Lusk and Briggeman (2009) used 11 food values, including price, taste, safety, and environmental impact as items in their BWS questions.

By highlighting the strength of its flexibility, BWS has been widely applied in various disciplines, for example:

- Food research
  - Beef (Lusk and Parker, 2009)
  - Pork (Jaeger et al., 2008)
  - Chocolate (Thomson et al., 2010)
  - Snacks (Mielby et al., 2012)
  - Breakfast bars (Hein et al., 2008)
  - Wine (Cohen, 2009, and many others)[1]
  - Juice (Jaeger and Cardello, 2009)
  - Coffee (Cohen and Neira, 2004)
  - Olive oil (Dekhili et al., 2011)
  - Restaurants (Chrzan and Golovashkina, 2006)
- Value research
  - Ethical beliefs (Auger et al., 2007)
  - Values (Lee et al., 2007, 2008)
  - Food safety issues (Finn and Louviere, 1992)
  - Food safety responsibility (Erdem et al., 2012)
  - Food values (Lusk and Briggeman, 2009)
  - Brand equity (Menictas et al., 2012)
  - Spirit of sport (Mazanov et al., 2012)
- Medical and health care research
  - Health care system reform (Louviere and Flynn, 2010)
  - Residency programs (Wang et al., 2011)
  - Treatment decisions in rheumatoid arthritis (van Hulst et al., 2011)
  - Measures for *Escherichia coli* O157 (Cross et al., 2012)
  - Side effects of smoking (Marti, 2012)
- Environmental and other research

[1]Some other wine studies include: Mueller and Rungie (2009); Goodman (2009); Cohen et al. (2009); Casini et al. (2009); Remaud and Lockshin (2009); Jaeger et al. (2009); de Magistris et al. (2011); Mueller Loose and Lockshin (2013); Sirieix et al. (2011); Bernabéu et al. (2012); Chrysochou et al. (2012a,b).

- Forest management programs (Kruger et al., 2013; Loureiro and Arcos, 2012)
- Scientists' opinions on biodiversity (Rudd, 2011)
- Conflict-handling style (Daly et al., 2010)
- Third-party logistics (Coltman et al., 2011)
- Business schools (Tavares et al., 2010)

This chapter, which explains how BWS is implemented in R, is organized as follows. After presenting the basic structure and features of BWS, Section 4.2 explains two methods for constructing choice sets for BWS, including an approach using a two-level orthogonal main-effect design and one using a balanced incomplete block design, as well as two methods for analyzing responses to BWS questions, namely, an approach counting the number of times each item is selected as the best and the worst, and one applying a conditional logit (CL) model. Section 4.3 describes how to implement these methods in R. Four packages are mainly used in this chapter: **DoE.base** (Grömping, 2014b), **crossdes** (Sailer, 2013), **support.BWS** (Aizaki, 2014a), and **survival** (Therneau, 2014). The first two packages provide functions for constructing choice sets for BWS, while the latter two provide those for constructing BWS questions, creating a dataset for analysis, and analyzing the responses to the BWS questions. Finally, Section 4.4 provides two examples of implementing BWS in R: a measure of the relative importance of the multifunctionality of agriculture and an evaluation of fruits by consumers.

4.2 Outline of BWS

4.2.1 BWS basics

Object case BWS is also known as Case 1 BWS or BW object scaling. After listing the items (objects) for evaluation by respondents, a number of different subsets of the items are constructed from the list according to the design of the experiment. Each of the subsets is presented as a choice set to the respondents, who are asked to select the best (or most important) item and the worst (or least important) item in the choice set. This question is repeated until all the subsets have been evaluated.

Figure 4.1 shows an example of BWS designed to measure citizens' preferences for the multifunctionality of agriculture, that is, a set of roles beyond the production of commodity outputs (see Section 4.4.1 for details of this example). According to the design of the experiment, various subsets of roles are selected from a list of nine multifunctional roles that have been determined in advance. Question 1 in this example illustrates one of the subsets comprising five roles: "biodiversity," "flood control," "food security," "animal welfare," and "cultural heritage." Respondents are asked to select the most and the least important roles from the five alternatives. In this example, a respondent selects "biodiversity" as the most important and "flood control" as the least important. Although only one question is illustrated in Figure 4.1,

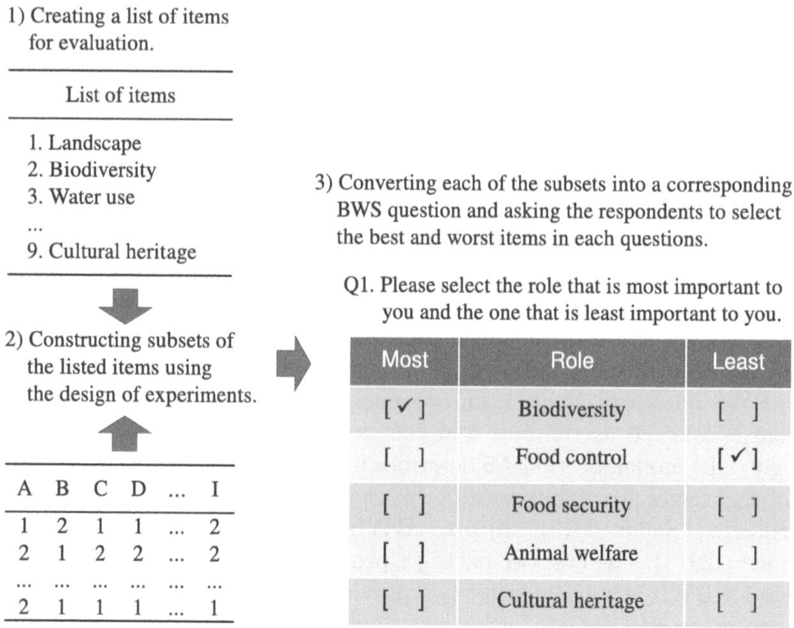

Figure 4.1 *An example of a BWS question.*

in the actual questionnaire this BWS question is repeated for all subsets of the roles. The number of questions is dependent on the number of items listed and the means by which the combinations of items are constructed.

There are three distinctive features of BWS application studies. The first is that comparative studies have been widely conducted. BWS calculates a unidimensional scale using the respondents' choice behaviors without requiring them to rate the items. Therefore, this scale avoids issues associated with rating scales (Cohen, 2003; Cohen and Neira, 2004). As an example, categories such as "very important" or "slightly important" are typically used as options in rating-scale questions without any guarantee that the interpretation of these options is the same for all respondents. However, there seems to be no room for discussion on the interpretation of selecting the best and worst items. As another example, best/worst selection behavior for various combinations of items does not enable us to select a specific option for all items (i.e., all items are evaluated as "very important"). As such, several international comparative studies have also been conducted (e.g., Auger et al., 2007; Cohen, 2009; Cohen et al., 2009; Goodman, 2009; Sirieix et al., 2011; Mueller Loose and Lockshin, 2013). The maximum number of countries in the above-mentioned studies is 12 (Goodman, 2009). Of course, there are also comparative studies that are based on respondents from only one country; these typically compare BWS scores for various respondents' categories classified by characteristics such as age (e.g. de Magistris et al., 2011; Chrysochou et al., 2012b).

The second feature is that BWS scores have been used as variables in other analyses. For example, Menictas et al. (2012) examined how brand equity constructs affect consumers' behavior when selecting a brand. Here, BWS was used to measure six brand equity constructs. The resultant BWS scores were then used as independent variables in a CL model that analyzed which of the three brands was selected by the respondents. Lee et al. (2008) measured BWS scores related to values, which are beliefs indicating what is important in a person's life, and then examined the relationships between the BWS scores and various attitudes and behaviors. Cohen et al. (2009) conducted a factor analysis of scores calculated from BWS questions related to factors affecting criteria for selecting wine in restaurants. The resultant factor loadings were then used to examine the differences in preferences for the criteria in three countries. Auger et al. (2007) classified respondents into groups by applying a cluster method to scores calculated from the results of BWS questions in which respondents were asked to select the most and least important social and ethical issues.

The final feature is that BWS has been compared with other methods that have been widely used for measuring the importance that individuals place on goods/services (e.g., Cohen, 2003; Chrzan and Golovashkina, 2006; Lee et al., 2007; Hein et al., 2008; Jaeger et al., 2008; Louviere and Islam, 2008; Mueller et al., 2009, 2010; Daly et al., 2010; Menictas et al., 2012; Mielby et al., 2012). For example, Chrzan and Golovashkina (2006) compared six methods (importance rating, constant sum, Q-sort, BWS, unbounded ratings, and magnitude estimation) from the viewpoint of task length, discriminative power, and predictive validity. They found that BWS has superior discriminative power and predictive validity. Menictas et al. (2012) compared BWS and confirmatory factor analysis for rating data from the viewpoint of measuring brand equity constructs. They found that BWS better explained individuals' behavior when selecting a brand. Mielby et al. (2012) compared ratings and BWS based on a survey targeting adolescent respondents from the viewpoint of the predictability of their actual choice of snacks. The respondents were asked to answer BWS questions by selecting their most and least preferred snacks, as well as rating the snacks on a hedonic scale. They then examined the relationship among BWS scores, the rating scale, and the actual choice of snacks. The results showed that a rating scale has relatively greater power when predicting the actual choice of snacks, whereas BWS has relatively greater power when discriminating between snacks.

4.2.2 Steps in BWS

The following provides the minimum information that is required to conduct BWS in R. Please refer to fundamental and comprehensive works for further details (e.g., Finn and Louviere, 1992; Marley and Louviere, 2005; Auger et al.,

Table 4.1 *An orthogonal design*

Row	A	B	C	D	E	F	G	H	I
1	1	2	1	1	2	1	2	2	2
2	2	1	2	2	2	1	1	1	2
3	1	2	1	2	2	2	1	1	1
4	2	1	1	2	1	2	2	2	1
5	2	2	2	1	1	1	2	1	1
6	1	1	1	2	1	1	2	1	2
7	2	2	2	2	2	2	2	2	2
8	2	2	1	1	1	2	1	1	2
9	1	1	2	1	2	2	2	1	1
10	1	1	2	1	1	2	1	2	2
11	1	2	2	2	1	1	1	2	1
12	2	1	1	1	2	1	1	2	1

2007; Flynn et al., 2007; Marley et al., 2008; Cohen, 2009; Flynn, 2010; Marley, 2010; Marley and Pihlens, 2012).

Step 1: Constructing choice sets

There are two methods for constructing choice sets for BWS: one uses a two-level orthogonal main-effect design (OMED) (Finn and Louviere, 1992), while the other uses a balanced incomplete block design (BIBD) (Auger et al., 2007).[2]

The first method uses a two-level OMED with m columns, where m is the total number of items evaluated: each column corresponds to an item and each row corresponds to a question. Each element is given one of the two distinct numbers in any two-level OMED; one of the values is interpreted as an item being "absent" from the corresponding column and the other as an item being "present." In this way, we can decide which items are assigned in each question.

For example, assume that $m = 9$ items (from item A to item I) are listed in a BWS study. Table 4.1 shows a two-level OMED with 9 columns, assigned to items A through I , respectively, generated using R (see Section 4.3.2 for the code to generate this design). Assuming that the value 1 denotes "absent" and the value 2 "present" in the design, the first row constructs a BWS question containing items B, E, G, H, and I (see Figure 4.2) since these five elements in the first row have the value 2, while the remaining elements have the value 1. Similarly, the second row in the design constructs a question containing the five items A, C, D, E, and I, and the last row in the design constructs

[2]Previous papers applying the two-level OMED method include Finn and Louviere (1992), Lusk and Briggeman (2009), Lusk and Parker (2009), Bernabéu et al. (2012), and Marti (2012), while those applying BIBD include Cohen and Neira (2004), Chrzan and Golovashkina (2006), Auger et al. (2007), Lee et al. (2007), Hein et al. (2008), and Jaeger et al. (2008).

Best	Item	Worst
[ ]	B	[ ]
[ ]	E	[ ]
[ ]	G	[ ]
[ ]	H	[ ]
[ ]	I	[ ]

Figure 4.2 *An example question in BWS based on the first row in Table 4.1.*

Table 4.2 *A balanced incomplete block design*

Block	1st item	2nd item	3rd item	4th item
1	1	4	6	7
2	2	3	4	6
3	2	4	5	7
4	1	2	5	6
5	1	3	4	5
6	3	5	6	7
7	1	2	3	7

a question containing the three items A, E, and H. It is easy to see that the BWS study based on this design comprises 12 questions since the number of rows is equal to the number of questions in BWS when using a two-level OMED and this design has 12 rows.

The second method uses a BIBD, which is a type of design in which a subset of treatments is assigned to each block (for the details see, e.g., Cochran and Cox, 1957; Morris, 2011). The features of a BIBD are given below:

- Number of treatments
- Size of a block
- Number of blocks
- Number of replications of each treatment
- Frequency with which each pair of treatments appears in the same block

When a BIBD is used to construct BWS questions, "treatment" and "block" in BIBD correspond, respectively, to "item" and "question" in BWS.

A feature of the BIBD approach is that the number of items per question (size of a question) is fixed to the number of columns in the BIBD, whereas the number of items per question varies according to the question in the two-level OMED approach.

For example, assume that the BWS questions are created using the BIBD approach for seven items, from item A to item G. Table 4.2 shows the BIBD

Best	Item	Worst
[ ]	A	[ ]
[ ]	D	[ ]
[ ]	F	[ ]
[ ]	G	[ ]

Figure 4.3 *An example question in BWS based on the first row in Table 4.2.*

that is available for 7 items, which is generated using R (see Section 4.3.2 for the code to generate this design). The BIBD has the following features: the number of items per question is 4 (4 columns), the number of questions is 7 (7 rows), the number of times each item is replicated in all the questions is 4, and the frequency with which each pair of items appears in the same question is 2. Based on the rule that a serial number between 1 and 7 is assigned to the items in alphabetical order (i.e., A = 1, B = 2, ..., G = 7), the question corresponding to the first row in the BIBD shows items A, D, F, and G (Figure 4.3). Similarly, the last row in the BIBD creates a question with the choice set containing items A, B, C, and G.

Despite the user having to search/generate 2-level OMEDs/BIBDs as required, a design satisfying all the conditions may not exist. Thus, the user will have to modify the conditions, such as the number of items, number of items per question, or number of questions. However, the number of items cannot be modified flexibly, since the aim of the study strongly affects the number of items. Therefore, modifying the number of items per question and/or the number of questions seems to be a realizable alternative. Here, we are interested in the desirable number of items per question and/or number of questions. Orme (2005) recommended four or five items per question based on a simulation study. Chrzan and Patterson (2006) found that the number of items per question increases the length of time needed by respondents to answer the BWS questions, and thus recommended a greater number of questions and smaller number of items per question. Figure 4.4 shows the number of items per question in BWS application papers using the BIBD approach, ranging from 3 to 8, with a mean and median of 4.4 and 4, respectively. The papers from which these statistics were derived are those cited in this chapter, and thus the result is not necessarily based on a comprehensive and systematic review of BWS application papers.[3]

[3]The number of items per question in those papers in which a two-level OMED was used to construct the choice sets ranged from 4 to 11.

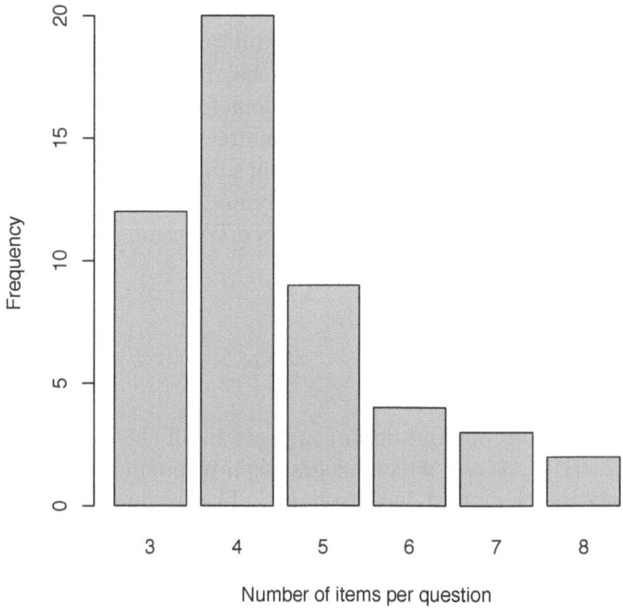

Figure 4.4 *Number of items per question taken from BWS papers using the BIBD approach.*

Step 2: Preparing a questionnaire survey

Survey design for BWS is basically similar to that for ordinal questionnaires. Guidelines for CV and DCE are also useful in preparing and conducting BWS surveys. Here, we discuss two aspects of conducting surveys for BWS: the division of choice sets and the sample size.

BWS questions are created by converting the choice sets constructed in the first step. Although choice sets are frequently divided into two or more subsets in the case of DCE studies (see Chapter 3), all choice sets are usually shown to each respondent in BWS studies. Of course, there are exceptions to this rule: studies that evaluate many items create different versions of BWS choice sets and assign each respondent to one of the different versions (e.g., van Hulst et al., 2011; Rudd, 2011).

From a statistical inference perspective, the sample size should be determined through sampling theory; when estimating characteristics of a population using BWS, the sample size can be calculated based on a formula considering multinomial proportion data (Louviere et al., 2013).

Step 3: Analyzing responses to questions

There are two approaches for analyzing responses to BWS questions: a counting approach and a modeling approach.

The counting approach calculates several types of scores based on the number of times (the frequency or count) that item i is selected as the best (B_{in}) and the worst (W_{in}) in all the questions for respondent n. Such scores are roughly divided into two categories: disaggregated (individual-level) scores and aggregated (total-level) scores (Finn and Louviere, 1992; Lee et al., 2007; Mueller and Rungie, 2009; Louviere and Flynn, 2010).

The first category includes a disaggregated BW score and its standardized score:

$$BW_{in} = B_{in} - W_{in}, \tag{4.1}$$

$$std.BW_{in} = \frac{BW_{in}}{r}, \tag{4.2}$$

where r is the number of times item i appears in all the questions. The maximum value of BW_{in} is $+r$, which occurs when respondent n selects item i as the best in all questions that include item i. The minimum value of BW_{in} is $-r$, which occurs when respondent n selects item i as the worst in all questions that include item i. The value of BW_{in} is zero when respondent n selects item i as the best with the same frequency as he/she selects item i as the worst, or when item i is selected as neither the best nor the worst. Since the standardized BW score ranges from -1 to $+1$, this is easy to understand.[4]

Aggregated scores (the second category of scores) are useful for interpreting trends in the responses of all respondents. The frequency with which item i is selected as the best in all the questions for all the respondents is denoted as B_i, and that with which item i is selected as the worst is denoted as W_i (i.e., $B_i = \sum_n B_{in}$, $W_i = \sum_n W_{in}$). The aggregated versions of BW_{in} and $std.BW_{in}$, as well as the square root of the ratio of B_i to W_i and its standardized score are defined as:

$$BW_i = B_i - W_i, \tag{4.3}$$

$$std.BW_i = \frac{BW_i}{Nr}, \tag{4.4}$$

$$sqrt.BW_i = \sqrt{\frac{B_i}{W_i}}, \tag{4.5}$$

$$std.sqrt.BW_i = \frac{sqrt.BW_i}{max.sqrt.BW_i}, \tag{4.6}$$

where N is the number of respondents and $max.sqrt.BW_i$ is the maximum value of $sqrt.BW_i$. The standardized $sqrt.BW_i$ helps us to understand the relative importance between items; thus, if $std.sqrt.BW_i$ is 0.5 and $std.sqrt.BW_j$

[4]There is another definition of the standardized BW score in which the right-hand side of Eq. (4.2) is multiplied by 100 (e.g., Goodman, 2009).

is 0.23, the resultant scale indicates that item i is approximately twice as important as item j ($0.5/0.23 \approx 2.2$).

The modeling approach uses discrete choice models to analyze responses and is based on understanding respondents' answers in the following situation (Finn and Louviere, 1992; Auger et al., 2007). Assume that m items exist in a choice set (a question). The number of possible pairs in which item i is selected as the best and item j is selected as the worst ($i \neq j$) from m items is $m \times (m-1)$. For each item, respondents are assumed to have a utility (v), which can be divided into two parts: the systematic and stochastic components. Assume that the respondents select items i and j as the best and worst, respectively, because the difference in utility between i and j represents the greatest utility difference. Furthermore, the stochastic components are assumed to be independent and identically distributed Gumbel variables. Under these assumptions, the probability of selecting item i as the best and item j as the worst is expressed as a CL model:[5]

$$Pr(i,j) = \frac{\exp(v_i - v_j)}{\sum_{k=1}^{m} \sum_{l=1, l \neq k}^{m} \exp(v_k - v_l)}. \tag{4.7}$$

It is worth noting that the denominator comprises a summation of all the differences in utility between items k and l ($l \neq k$). If the number of items per question is four, the number of possible best–worst pairs of two items from four items is twelve; thus, 12 utility pairs have to be calculated. This number differs from that in a DCE under the same condition of four items (alternatives) in a choice set.

When estimating the CL model formulated above, the utility for an arbitrary item is normalized; that is, the coefficient of the normalized item is fixed at zero. Therefore, other coefficients show the difference in value from the coefficient of the normalized item.

Each estimated utility (coefficient) is frequently converted into a "share of preference" for item i (SP_i) based on the CL model choice rule (Cohen, 2003; Cohen and Neira, 2004; Lusk and Briggeman, 2009):

$$SP_i = \frac{\exp(v_i)}{\sum_{j=1}^{m} \exp(v_j)}. \tag{4.8}$$

This scale shows the relative importance between items as reflected by the standardized $sqrt.BW_i$.

CL estimates are provided as average values across all respondents, while count-based scores are calculated for each respondent. Therefore, to conduct further empirical analysis, some studies use the counting approach. A benefit of the modeling approach is the ability to consider whether the average

[5]The following discrete choice models have also been used: mixed logit model (e.g., Lusk and Briggeman, 2009; Cross et al., 2012; Marti, 2012), latent class choice model (e.g., Cohen and Neira, 2004; Mueller Loose and Lockshin, 2013), and the hierarchical Bayesian choice model (e.g., Chrzan and Golovashkina, 2006; Rudd, 2011; van Hulst et al., 2011).

response follows the theoretical background on which respondents' choice behaviors are assumed to be based. Another benefit is the ability to test whether each coefficient differs statistically from zero at the same time as estimating the model. Of course, such a statistical test is also available in the counting approach (e.g., Cohen, 2009).

4.3 R functions for BWS

4.3.1 Overview

Figure 4.5 illustrates the functions used in each step of BWS. In the first step (constructing choice sets), R functions related to the design of the experiment are available to generate a two-level OMED and BIBD: the `oa.design()` function in the **DoE.base** package generates a two-level OMED, while the `find.BIB()` function in the **crossdes** package generates a BIBD. In the second step (preparing a questionnaire), the `bws.questionnaire()` function in the **support.BWS** package can be used to convert the choice sets into BWS questions. In the last step (analyzing responses), the `bws.dataset()` function in **support.BWS** is used to create a dataset for analysis by combining the design created in the first step and the dataset containing the responses to the BWS questions. Functions `bws.count()` in **support.BWS** and `clogit()` in **survival** can be used to analyze the responses to BWS using the counting and modeling approaches, respectively.

In preparation for implementing the example code using these functions given in Sections 4.3 and 4.4, we start by loading the four packages into R:

```
R> library(DoE.base)
R> library(crossdes)
R> library(support.BWS)
R> library(survival)
```

4.3.2 Constructing choice sets

Two-level OMEDs can be generated using the `oa.design()` function in the **DoE.base** package. A generic call to the function for the purpose of generating a two-level OMED used in a BWS study is shown below:[6]

```
oa.design(nfactors,
          nlevels)
```

with arguments:

- `nfactors`: Number of factors (items).
- `nlevels`: Level of factors.

On the other hand, BIBDs can be generated using the `find.BIB()` function in the **crossdes** package. A generic call to the function is illustrated below.

[6]Function `oa.design()` generates various orthogonal designs, and has other arguments besides `nfactors` and `nlevels`. Consult the application function help facility for the details.

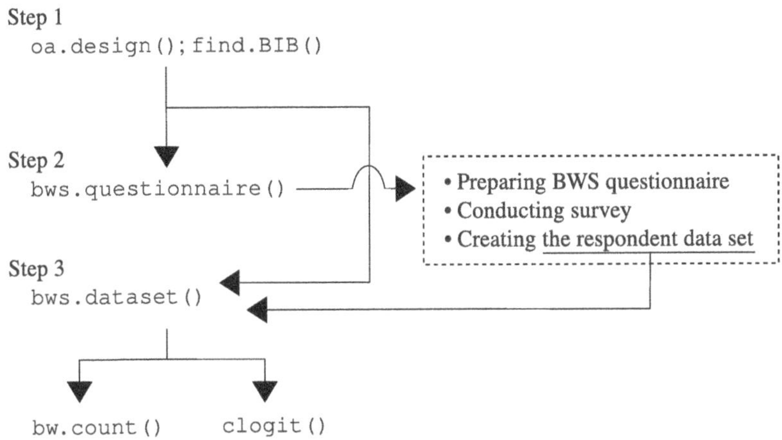

Figure 4.5 *Functions used in each step of BWS.*

```
find.BIB(trt,
         k,
         b)
```

with arguments:

- `trt`: Number of treatments (items).
- `k`: Number of items per question.
- `b`: Number of questions.

Note that we need to check whether the resultant design is a BIBD by using the `isGYD()` function in the same package.

If these functions do not generate the required design, the reader is advised to consult the many books (e.g., Cochran and Cox, 1957; Hedayat et al., 1999) and/or websites (e.g., Kuhfeld, 2006; Sloane, 2014) related to the design of experiments. However, there is the possibility that the design that satisfies the required conditions may not exist. In this case, the conditions of the study need to be modified.[7]

Example code and output

As an example of the `oa.design()` function, the following code returns an orthogonal design with nine factors in two levels:

```
R> set.seed(123)
R> oa <- oa.design(nfactors = 9, nlevels = 2)
```

[7]In addition to these functions, various functions related to the design of experiments in R have been developed. Please refer to *CRAN's Task View: Design of Experiments (DoE) & Analysis of Experimental Data* (Grömping, 2014a) for the details.

```
The columns of the array have been used in order of appearance.
For designs with relatively few columns,
the properties can sometimes be substantially improved
using option columns with min3 or even min34.

R> oa
   A B C D E F G H J
1  1 2 1 1 2 1 2 2 2
2  2 1 2 2 2 1 1 1 2
3  1 2 1 2 2 2 1 1 1
4  2 1 1 2 1 2 2 2 1
5  2 2 2 1 1 1 2 1 1
6  1 1 1 2 1 1 2 1 2
7  2 2 2 2 2 2 2 2 2
8  2 2 1 1 1 2 1 1 2
9  1 1 2 1 2 2 2 1 1
10 1 1 2 1 1 2 1 2 2
11 1 2 2 2 1 1 1 2 1
12 2 1 1 1 2 1 1 2 1
class=design, type= oa
```

As seen in the above example, messages are frequently displayed immediately after executing the function if it works correctly. For a detailed explanation of these messages, refer to the help for `oa.design()`.

As an example of a BIBD, the following conditions are assumed:

- Number of items is seven.
- Number of items per question is four.
- Number of questions is seven.

The code corresponding to the above is given below:

```
R> set.seed(123)
R> bibd <- find.BIB(trt = 7, k = 4, b = 7)
R> bibd
     [,1] [,2] [,3] [,4]
[1,]    1    4    6    7
[2,]    2    3    4    6
[3,]    2    4    5    7
[4,]    1    2    5    6
[5,]    1    3    4    5
[6,]    3    5    6    7
[7,]    1    2    3    7
```

while the code for checking whether the resultant design `bibd` is a BIBD is:

```
R> isGYD(bibd)
[1] The design is a balanced incomplete block design w.r.t. rows.
```

Tables 4.1 and 4.2 are generated using these two code examples, respectively. Note that the resultant designs may differ from these when other values are assigned to the argument of `set.seed()` or when `set.seed(123)` is not executed.

4.3.3 Preparing BWS questions

The `bws.questionnaire()` function in the **support.BWS** package is used to convert the two-level OMED/BIBD into a series of BWS questions. A generic call to the function is:

```
bws.questionnaire(choice.sets,
                  design.type,
                  item.names)
```

with its arguments:

- `choice.sets`: A data frame or matrix containing choice sets.
- `design.type`: A value describing how to design the choice sets, which is set to `1` if the design assigned to the argument `choice.sets` is a two-level OMED, and `2` if it is a BIBD.
- `item.names`: A vector containing the names of items shown in the questions.

The design is assigned to the argument `choice.sets`, and may be generated by the functions mentioned above or copied manually from text books/websites related to the design of experiments. If the design is a two-level OMED, each row corresponds to a question and each column corresponds to an item. The level values must be 1 and 2, where the former corresponds to an item being "absent" from a column and the latter corresponds to the item being "present." The correspondence between items and columns is defined and assigned to the argument `item.names`, where the order of names in the vector assigned to `item.names` must correspond with the order of columns in the choice sets assigned to `choice.sets`.

If the design is a BIBD, each row corresponds to a question and the number of columns is equal to the number of items per question. The level values must be serial integer values, starting from 1, with each value corresponding to an item. The correspondence between items and level values is defined and assigned to the argument `item.names`, where the order of names in `item.names` must correspond to the order of the level values in the design (i.e., the j-th item in `item.names` corresponds to level value j in the design).

Example code and output

As an example of creating BWS questions, the following conditions are assumed:

- BWS questions are created using the BIBD generated in Section 4.3.2.
- Items are named A, B, C, D, E, F, and G.

- Level values in the BIBD ranging from 1 to 7 are assigned to the items in alphabetical order (i.e., A = 1, B = 2, ..., G = 7).

Given these assumptions, the names of the items are defined in vector `items` as follows:

```
R> items <- c("A", "B", "C", "D", "E", "F", "G")
```

Then, a set of seven BWS questions based on `bibd` and `items` is displayed using the `bws.questionnaire()` function:

```
R> bws.questionnaire(choice.sets = bibd,
                     design.type = 2,  # BIBD
                     item.names = items)

Q1
 Best Items Worst
 [ ]  A     [ ]
 [ ]  D     [ ]
 [ ]  F     [ ]
 [ ]  G     [ ]

Q2
 Best Items Worst
 [ ]  B     [ ]
 [ ]  C     [ ]
 [ ]  D     [ ]
 [ ]  F     [ ]

...

Q7
 Best Items Worst
 [ ]  A     [ ]
 [ ]  B     [ ]
 [ ]  C     [ ]
 [ ]  G     [ ]
```

We can copy the resultant questions from the R console, paste them into other software packages such as a word processor, and then modify them to create the questionnaire.

4.3.4 Creating the dataset

The `bws.dataset()` function in the **support.BWS** package is available for creating a dataset suitable for analysis using the counting and modeling approaches. A generic call to the function is:

```
bws.dataset(respondent.data,
            response.type = 1,
```

```
              choice.sets,
              design.type = 1,
              item.names = NULL,
              row.renames = TRUE)
```

with its arguments:

- `respondent.data`: A data frame containing the respondent dataset.
- `response.type`: A value describing the format of the response variables, which is set to `1` if the response variables follow row number format, and `2` if they follow item number format.
- `choice.sets`: A data frame or matrix containing choice sets.
- `design.type`: A value describing how to design the choice sets, which is set to `1` if the design assigned to the argument `choice.sets` is a two-level OMED, and `2` if it is a BIBD.
- `item.names`: A vector containing the names of items or `NULL` (default), in which case default item names (i.e., `ITEM1`, `ITEM2`, ...) are used in the resultant dataset.
- `row.renames`: A logical variable describing whether or not the row names of a dataset created by this function are changed. If `TRUE`, integer values are assigned to the row names starting from `1`. If `FALSE`, the row names are not changed.

The respondent dataset, in which each row corresponds to a respondent, must be organized by the users. The dataset must contain the ID variable in the first column, denoting the respondent's identification number, and pairs of response variables in subsequent columns, with each pair indicating which items are selected as the best and the worst for each question. Although the names of the ID and response variables are left to the discretion of the users, the response variables must be ordered in such a way that the best option alternates with the worst for each question. For example, if there are seven BWS questions, the variables are `B1 W1 B2 W2 ... B7 W7`, where `Bi` and `Wi` depict which items are selected as the best and the worst in the `i`-th question.

In addition, there are two types of data format related to the response variables: one is row number format, and the other is item number format. In the former, the row numbers of the items that are selected as the best and the worst are stored in the response variables. In the latter, item numbers are stored in the response variables. For example, consider the BWS question shown in Figure 4.3, which corresponds to the first question in Table 4.2. Assume that the respondent selects item G as the best and item A as the worst for this question. When using row number format, variables `B1` and `W1` for the respondent have values `4` and `1`, respectively, because items G and A are located in the 4th and 1st rows in the question, respectively. When using item number format, variables `B1` and `W1` for the respondent have values `7` and `1`, respectively, because items G and A are numbered as 7 and 1, respectively.

Table 4.3 *An example of an artificial respondent dataset (N = 10)*

ID	B1	W1	B2	W2	B3	W3	B4	W4	B5	W5	B6	W6	B7	W7
1	2	3	4	3	3	1	2	4	1	3	2	3	2	4
2	1	4	3	2	1	4	2	4	3	1	1	2	1	4
3	2	3	3	4	1	2	1	3	2	4	1	3	3	4
4	4	3	3	1	1	2	3	4	1	4	3	4	1	4
5	2	1	3	2	1	4	2	1	3	1	2	3	3	4
6	2	3	1	3	1	4	2	3	2	4	4	2	2	1
7	2	3	1	3	1	2	2	4	1	3	4	2	3	4
8	1	2	2	4	1	3	2	1	1	2	3	4	3	1
9	2	3	1	2	2	3	4	2	1	4	3	4	2	4
10	1	3	1	4	1	3	1	4	1	3	3	2	2	4

The arguments `choice.sets`, `design.type`, and `item.names` are the same as those in the `bws.questionnaire()` function. However, `item.names` can take `NULL`, in which case default item names (i.e., `ITEM1`, `ITEM2`, ...) are used in the resultant dataset. Further note that the order of the questions in the choice sets must be the same as that in the respondent dataset.

Example code and output

The `bws.dataset()` function provides a data frame in the format where one row depicts one possible pair of item i selected as the best and item j selected as the worst ($i \neq j$), because the difference in utility/value between the two items i and j is a fundamental calculation as explained in Section 4.2.2.

Table 4.3 shows an artificial respondent dataset in row number format, which contains ten respondents' answers to BWS questions created from the BIBD in Section 4.3.2. We assume that each respondent is asked to answer seven BWS questions; therefore, the variables containing their responses each contain seven pairs of `Bi` and `Wi`: (`B1 W1 B2 W2 ... B7 W7`). Since there are four items in each question and the response variables are prepared in row number format, the values of `Bi` and `Wi` range from `1` to `4`, where `1`, `2`, `3`, and `4` denote an item shown in the 1st, 2nd, 3rd, and 4th row of the BWS question, respectively. Accordingly, the answers to the BWS questions can be interpreted as follows. Respondent 1 selected an item in the 2nd row as the best and an item in the 3rd row as the worst in question 1 (according to the values of `B1` and `W1` in the row with `ID = 1`). Respondent 10 selected an item in the 2nd row as the best and an item in the 4th row as the worst in question 7 (according to the values of `B7` and `W7` in the row with `ID = 10`).

The following code assigns the respondent dataset shown in Table 4.3 to an object `res`, creates the dataset, which is based on the `bibd` created in Section 4.3.2 and the respondent dataset `res`, and then shows part of the dataset shown in Figure 4.6:

```
R> res <- data.frame( # row number format
```

```
        ID = c(1, 2, 3, 4, 5, 6, 7, 8, 9, 10),
        B1 = c(2, 1, 2, 4, 2, 2, 2, 1, 2, 1),
        W1 = c(3, 4, 3, 3, 1, 3, 3, 2, 3, 3),
        B2 = c(4, 3, 3, 3, 3, 1, 1, 2, 1, 1),
        W2 = c(3, 2, 4, 1, 2, 3, 3, 4, 2, 4),
        B3 = c(3, 1, 1, 1, 1, 1, 1, 1, 2, 1),
        W3 = c(1, 4, 2, 2, 4, 4, 2, 3, 3, 3),
        B4 = c(2, 2, 1, 3, 2, 2, 2, 2, 4, 1),
        W4 = c(4, 4, 3, 4, 1, 3, 4, 1, 2, 4),
        B5 = c(1, 3, 2, 1, 3, 2, 1, 1, 1, 1),
        W5 = c(3, 1, 4, 4, 1, 4, 3, 2, 4, 3),
        B6 = c(2, 1, 1, 3, 2, 4, 4, 3, 3, 3),
        W6 = c(3, 2, 3, 4, 3, 2, 2, 4, 4, 2),
        B7 = c(2, 1, 3, 1, 3, 2, 3, 3, 2, 2),
        W7 = c(4, 4, 4, 4, 4, 1, 4, 1, 4, 4))
R> dat <- bws.dataset(respondent.dataset = res,
                      response.type = 1,  # row number format
                      choice.sets = bibd,
                      design.type = 2,    # BIBD
                      item.names = items)

R> dat[1:85,]
```

The variables contained in the dataset generated by the function are explained below:

- `Q`: A serial number for the BWS questions starting at `1`.
- `PAIR`: A serial number of the possible pairs of the best and worst items for each question.
- `ID`: A respondent's identification number, assigned according to the respondent identification variable in the respondent dataset.
- `RES.B`: Item numbers selected as the best by respondents.
- `RES.W`: Item numbers selected as the worst by respondents.
- `BEST`: Item numbers treated as the best in the possible pairs of the best and worst items for each question.
- `WORST`: Item numbers treated as the worst in the possible pairs of the best and worst items for each question.
- `ITEMj`: State variable related to the possible pairs of the best and worst items; it takes a value of `1` if item `j` is treated as the best in the possible pair, a value of `-1` if item `j` is treated as the worst in the possible pair, and a value of `0` otherwise. If the vector containing the names of items is assigned to the argument `item.names`, the assigned names are used instead of `ITEMj`.
- `RES`: Response to a BWS question, which is set to `TRUE` if a possible pair of the best and worst items is selected by the respondents, and `FALSE`

The respondent ID = 1 selected 4th item (D) as best (RES.B = 4) and 6th item (F) as worst (RES.W = 6) in question Q = 1.

Since a possible pair corresponding to row 1 is that A is treated as best (BEST = 1) and D is treated as worst (WORST = 4), A = 1 and D = −1 in the row.

Twelve possible pairs of best item and worst item in question Q = 1 for the respondent ID = 1 are located between row 1 and row 12.

Q	PAIR	ID	RES.B	RES.W	BEST	WORST	A	B	C	D	E	F	G	RES	STR
1	1	1	4	6	1	4	1	0	0	-1	0	0	0	FALSE	101
1	2	1	4	6	1	6	1	0	0	0	0	-1	0	FALSE	101
1	3	1	4	6	1	7	1	0	0	0	0	0	-1	FALSE	101
1	4	1	4	6	4	1	-1	0	0	1	0	0	0	FALSE	101
1	5	1	4	6	4	6	0	0	0	1	0	-1	0	TRUE	101
1	6	1	4	6	4	7	0	0	0	1	0	0	-1	FALSE	101
1	7	1	4	6	6	1	-1	0	0	0	0	1	0	FALSE	101
1	8	1	4	6	6	4	0	0	0	-1	0	1	0	FALSE	101
1	9	1	4	6	6	7	0	0	0	0	0	1	-1	FALSE	101
1	10	1	4	6	7	1	-1	0	0	0	0	0	1	FALSE	101
1	11	1	4	6	7	4	0	0	0	-1	0	0	1	FALSE	101
1	12	1	4	6	7	6	0	0	0	0	0	-1	1	FALSE	101
2	1	1	6	4	2	3	0	1	-1	0	0	0	0	FALSE	102
2	2	1	6	4	2	4	0	1	0	-1	0	0	0	FALSE	102
...	...	...	...	...	...	...	...	...	...	...	...	...	...	...	...
2	12	1	6	4	6	4	0	0	0	-1	0	1	0	TRUE	102
3	1	1	5	2	2	4	0	1	0	-1	0	0	0	FALSE	103
...	...	...	...	...	...	...	...	...	...	...	...	...	...	...	...
7	12	1	2	7	7	3	0	0	-1	0	0	0	1	FALSE	107
1	1	2	1	7	1	4	1	0	0	-1	0	0	0	FALSE	201

These are the possible pairs in question Q = 1 for the respondent ID = 1.

RES = TRUE at a row corresponding to BEST = 4 and WORST = 6 in question Q = 1 for the respondent ID = 1.

Figure 4.6 *Example output from function* **bws.dataset()**.

otherwise. This variable is used as a dependent variable in model `formula` of the `clogit()` function in the **survival** package when analyzing responses to BWS questions (a more detailed explanation is provided later).

- `STR`: A stratification variable identifying each combination of respondent and question. This variable is also used in model `formula` of `clogit()`.

The BIBD contains seven treatments (seven items) and consists of seven blocks of size four (seven questions containing four items each). Therefore, the number of possible pairs of the best and worst items for each question is twelve ($= 4 \times (4-1)$). The twelve possible pairs for question `Q = 1` and the respondent with `ID = 1` are given by the variables `BEST` and `WORST` for each of rows 1 to 12. Row 1 shows the possible pair representing the case where A is considered the best (`BEST = 1`) and D is considered the worst (`WORST = 4`); therefore, variable `A` has the value `1`, variable `D` has the value `-1`, and the other state variables `B`, `C`, `E`, `F`, and `G` have the value `0` in row 1. We can see that the respondent with `ID = 1` selected the 4th item (`D`) as the best and the 6th item (`F`) as the worst in question `Q = 1` from variables `RES.B` and `RES.W` for rows 1 to 12. A possible pair corresponding to `BEST = 4` and `WORST = 6` in question `Q = 1` for the respondent with `ID = 1` is located in row 5; therefore, variable `RES` has `TRUE` in row 5 and `FALSE` for all rows between 1 and 12, except row 5.

For the sake of completeness, choice sets in item number format that are converted from `res` are given as:

```
R> res2 <- data.frame( # item number format
     ID = c(1, 2, 3, 4, 5, 6, 7, 8, 9, 10),
     B1 = c(4, 1, 4, 7, 4, 4, 4, 1, 4, 1),
     W1 = c(6, 7, 6, 6, 1, 6, 6, 4, 6, 6),
     B2 = c(6, 4, 4, 4, 4, 2, 2, 3, 2, 2),
     W2 = c(4, 3, 6, 2, 3, 4, 4, 6, 3, 6),
     B3 = c(5, 2, 2, 2, 2, 2, 2, 2, 4, 2),
     W3 = c(2, 7, 4, 4, 7, 7, 4, 5, 5, 5),
     B4 = c(2, 2, 1, 5, 2, 2, 2, 2, 6, 1),
     W4 = c(6, 6, 5, 6, 1, 5, 6, 1, 2, 6),
     B5 = c(1, 4, 3, 1, 4, 3, 1, 1, 1, 1),
     W5 = c(4, 1, 5, 5, 1, 5, 4, 3, 5, 4),
     B6 = c(5, 3, 3, 6, 5, 7, 7, 6, 6, 6),
     W6 = c(6, 5, 6, 7, 6, 5, 5, 7, 7, 5),
     B7 = c(2, 1, 3, 1, 3, 2, 3, 3, 2, 2),
     W7 = c(7, 7, 7, 7, 7, 1, 7, 1, 7, 7))
```

It is self-evident that the dataset `dat2` constructed from `res2` is the same as `dat`, which is constructed from `res`:

```
R> dat2 <- bws.dataset(respondent.dataset = res2,
                       response.type = 2,  # item number format
                       choice.sets = bibd,
                       design.type = 2,    # BIBD
```

```
                              item.names = items)
R> identical(dat, dat2) # check whether dat is identical to dat2
[1] TRUE
```

4.3.5 Analyzing responses using the counting approach

To calculate count-based scores, function `bws.count()` in package **support.BWS** is available. A generic call to the function and definition of its arguments are given below:

```
bws.count(data)
```

- `data`: A dataset generated by function `bws.dataset()`.

Example code and output

The following example code calculates BWS scores based on the respondent dataset created in Section 4.3.4:

```
R> scores <- bws.count(data = dat)
R> scores
Number of respondents = 10

Summary of disaggregated best-worst scores:
  meanB meanW meanBW mean.stdBW stdev.stdBW
A   1.3   0.7    0.6      0.150       0.412
B   2.2   0.3    1.9      0.475       0.299
C   0.9   0.4    0.5      0.125       0.270
D   1.3   1.0    0.3      0.075       0.409
E   0.4   1.3   -0.9     -0.225       0.381
F   0.6   1.8   -1.2     -0.300       0.284
G   0.3   1.5   -1.2     -0.300       0.230

Aggregated best-worst scores:
   B  W  BW  stdBW sqrtBW std.sqrtBW
A 13  7   6  0.150  1.363      0.503
B 22  3  19  0.475  2.708      1.000
C  9  4   5  0.125  1.500      0.554
D 13 10   3  0.075  1.140      0.421
E  4 13  -9 -0.225  0.555      0.205
F  6 18 -12 -0.300  0.577      0.213
G  3 15 -12 -0.300  0.447      0.165
```

Since the output from `bws.count()` is an object of the S3 class "`bws.count`" and a `print()` function for this class has been defined, formatted output is returned as shown above. The "summary of disaggregated best–worst scores" table shows the means of disaggregated B, W, and BW, and the mean and standard deviation of the standardized disaggregated BW scores by item. The "aggregated best–worst scores" table shows the aggregated B, W, BW, and standardized BW scores by item, and the square root of the ratio of B and W and its standardized scores by item.

Each of the components in the output from `bws.count()` is explained below:

- The list `disaggregate` contains five objects related to disaggregated (individual-level) scores.
  - `ID`: A vector giving the respondent's identification number.
  - `B`: A matrix giving the number of times item i is selected as the best by each respondent.
  - `W`: A matrix giving the number of times item i is selected as the worst by each respondent.
  - `BW`: A matrix giving the difference between `B` and `W` for item i per respondent.
  - `stdBW`: A matrix giving the standardized `BW`.
- The data frame `aggregate` contains the aggregated (total-level) scores for all respondents.
  - `B`: A variable giving the number of times item i is selected as the best for all respondents.
  - `W`: A variable giving the number of times item i is selected as the worst for all respondents.
  - `BW`: A variable giving the difference between `B` and `W` for item i for all respondents.
  - `stdBW`: A variable giving the standardized `BW`.
  - `sqrtBW`: A variable giving the square root of the ratio of `B` to `W` for item i for all respondents.
  - `std.sqrtBW`: A variable giving the standardized `sqrtBW`.
- The list `information` contains basic information related to the BWS questions.
  - `nrespondents`: A variable giving the number of respondents.
  - `nitems`: A variable giving the number of items.
  - `fitem`: A variable giving the frequency of each item in the choice sets.
  - `vnames`: A variable giving the names of each item.

Individual objects in the output are accessible using the `$` operator. For example, the following code returns `BW` in `aggregate`:

```
R> scores$aggregate$BW
[1]   6  19   5   3  -9 -12 -12
```

4.3.6 Analyzing responses using the modeling approach

A CL model analysis of responses to BWS questions can be conducted using the `clogit()` function in the **survival** package. A generic call to the function and its arguments for the purpose of applying a CL model to the BWS responses are given as:

```
clogit(formula,
       data)
```

- `formula`: The model formula to be estimated.
- `data`: A data frame containing the dataset created by function `bws.dataset()`.

When applying `clogit()` to BWS questions in which a total of j items are evaluated by the respondents, a typical structure of the model `formula` is defined as:

```
RES ~ ITEM1 + ITEM2 + ... + ITEMj-1 + strata(STR)
```

where the variables are the same as those defined in Section 4.3.4. The j-th item (variable `ITEMj`) is excluded from the `formula` because the coefficient of an arbitrary item must be normalized to zero. Therefore, other coefficients show the difference in utility from the j-th item. Here, `strata(STR)` denotes that variable `STR` is used to identify each combination of respondent and BWS question.

Example code and output

The following code can be used to carry out CL model analysis based on the respondent dataset (`dat`) created in Section 4.3.4 under the assumption that the coefficient of item `D` is normalized to zero:

```
R> fr <- RES ~ A + B + C + E + F + G + strata(STR)
R> clogit(formula = fr, data = dat)
Call:
clogit(formula = fr, data = dat)


     coef exp(coef) se(coef)      z     p
A  0.1438     1.155    0.310  0.464 0.640
B  0.7910     2.205    0.324  2.441 0.015
C  0.0753     1.078    0.316  0.239 0.810
E -0.5907     0.554    0.318 -1.857 0.063
F -0.7360     0.479    0.319 -2.306 0.021
G -0.7334     0.480    0.322 -2.278 0.023

Likelihood ratio test=36.6  on 6 df, p=2.11e-06  n= 840,
number of events= 70
```

Estimates are shown in table format, where each row corresponds to an independent variable (`A`, `B`, ..., `G`), and columns `coef` to `p` show the estimated coefficient, exponential transformation of the estimate, standard error, z-value, and p-value, respectively. Note that since the coefficient of item D is normalized to zero, other coefficients show the difference in value from the coefficient of item D. According to columns `coef` and `p`, item `B` is more important than item `D`; items `E`, `F`, and `G` are less important than `D`; and items `A` and `C` are indifferent to item `D`. Consult the application help function for `clogit()` for details of the output.

4.4 Example BWS using R

This section provides two examples showing how the functions discussed above can be used in BWS: a measure of the relative importance of the multifunctionality of agriculture and an evaluation of fruits by consumers.

4.4.1 BWS based on a two-level OMED

The BWS example using a two-level OMED is a hypothetical study that elicits citizens' preferences for the multifunctionality of agriculture. Agriculture produces both multiple commodities and non-commodity outputs, with some of the latter having characteristics of externalities; in other words, although citizens obtain benefits from these non-commodity outputs, they do not pay for them because of the lack of markets for such outputs. Eliciting citizens' preferences for such multifunctionality using various methods is a major research topic in agricultural economics.[8]

This example of a BWS application targets nine roles:

- Landscape
- Biodiversity
- Water use
- Land conservation
- Flood control
- Rural viability
- Food security
- Animal welfare
- Cultural heritage

A two-level OMED with nine columns is generated and stored in the matrix `sets.mfa`:

```
R> sets.mfa <- cbind( # 12 rows x 9 columns
     c(1, 2, 1, 2, 2, 1, 2, 2, 1, 1, 1, 2), # 1st column
     c(2, 1, 2, 1, 2, 1, 2, 2, 1, 1, 2, 1), # 2nd column
```

[8]Reviews related to this multifunctionality include OECD (2001), Randall (2002), Madureira et al. (2007), Bergstrom and Ready (2009), and Renting et al. (2009).

```
    c(1, 2, 1, 1, 2, 1, 2, 1, 2, 2, 2, 1), # 3rd column
    c(1, 2, 2, 2, 1, 2, 2, 1, 1, 1, 2, 1), # 4th column
    c(2, 2, 2, 1, 1, 1, 2, 1, 2, 1, 1, 2), # 5th column
    c(1, 1, 2, 2, 1, 1, 2, 2, 2, 2, 1, 1), # 6th column
    c(2, 1, 1, 2, 2, 2, 2, 1, 2, 1, 1, 1), # 7th column
    c(2, 1, 1, 2, 1, 1, 2, 1, 1, 2, 2, 2), # 8th column
    c(2, 2, 1, 1, 1, 2, 2, 2, 1, 2, 1, 1)) # 9th column
R> sets.mfa
      [,1] [,2] [,3] [,4] [,5] [,6] [,7] [,8] [,9]
 [1,]    1    2    1    1    2    1    2    2    2
 [2,]    2    1    2    2    2    1    1    1    2
 [3,]    1    2    1    2    2    2    1    1    1
 [4,]    2    1    1    2    1    2    2    2    1
 [5,]    2    2    2    1    1    1    2    1    1
 [6,]    1    1    1    2    1    1    2    1    2
 [7,]    2    2    2    2    2    2    2    2    2
 [8,]    2    2    1    1    1    2    1    1    2
 [9,]    1    1    2    1    2    2    2    1    1
[10,]    1    1    2    1    1    2    1    2    2
[11,]    1    2    2    2    1    1    1    2    1
[12,]    2    1    1    1    2    1    1    2    1
```

This design is the same as that in Table 4.1, where each row corresponds to a question, and each column corresponds to an item. A level value of `1` (2) denotes an item being "absent (present)" from the corresponding column.[9]

The correspondence between items (multifunctional roles) and columns is defined and stored in the vector `items.mfa`:

```
R> items.mfa <- c(
    "Landscape",          # 1st column in sets.mfa
    "Biodiversity",       # 2nd column in sets.mfa
    "Water use",          # 3rd column in sets.mfa
    "Land conservation",  # 4th column in sets.mfa
    "Flood control",      # 5th column in sets.mfa
    "Rural viability",    # 6th column in sets.mfa
    "Food security",      # 7th column in sets.mfa
    "Animal welfare",     # 8th column in sets.mfa
    "Cultural heritage")  # 9th column in sets.mfa
```

This example uses BWS questions formatted in the same way as that in Figure 4.1. By using function `bws.questionnaire()`, with arguments relevant to this example, the following twelve BWS questions are created:

```
R> bws.questionnaire(choice.sets = sets.mfa,
                     design.type = 1, # OMED
                     item.names = items.mfa)
```

[9]See Section 4.3.2 for the code to generate this design.

```
Q1
 Best Items              Worst
 [ ]  Biodiversity       [ ]
 [ ]  Flood control      [ ]
 [ ]  Food security      [ ]
 [ ]  Animal welfare     [ ]
 [ ]  Cultural heritage [ ]

...

Q12
 Best Items          Worst
 [ ]  Landscape      [ ]
 [ ]  Flood control  [ ]
 [ ]  Animal welfare [ ]
```

Assume that 100 individuals answered all twelve BWS questions, creating a respondent dataset that is included in the `mfa` dataset in the **support.BWS** package. The dataset can be loaded as follows:

```
R> data(mfa)
R> head(mfa) # display first 6 rows
  ID B1 W1 B2 W2 B3 W3 B4 W4 B5 W5 B6 W6 B7 W7 B8 W8 B9 W9
1  1  5  4  1  5  1  3  2  4  2  4  2  1  2  7  2  3  3  1
2  2  4  2  5  4  2  3  3  1  3  4  1  2  3  6  1  3  4  2
3  3  1  3  5  4  1  3  3  5  2  3  3  2  1  8  2  3  3  2
4  4  5  3  5  1  1  3  5  3  2  4  3  2  2  5  2  1  1  3
5  5  5  2  1  3  4  3  5  3  2  3  3  1  1  3  4  3  4  3
6  6  5  1  5  2  4  3  3  5  1  3  3  2  6  2  2  3  2  3
  B10 W10 B11 W11 B12 W12
1   3   1   3   2   3   2
2   4   1   1   4   3   1
3   4   1   1   2   1   2
4   3   1   1   2   1   3
5   4   1   3   1   1   2
6   2   1   1   2   1   2
```

The first column is the variable `ID`, which gives the respondent's identification number. The remaining columns are response variables for the BWS questions. Since there are twelve questions in this example, the dataset contains twelve pairs of `Bi` and `Wi`; for example, `B1` and `W1` correspond to the 1st question, while `B12` and `W12` correspond to the 12th question. The data format is assumed to be row number format in the dataset. For example, the response variables `B1` and `W1` for the respondent with `ID = 1` contain `5` and `4`, respectively. Since the 5th and 4th rows in question 1 denote cultural heritage and animal welfare, respectively, this means that the respondent selected cultural heritage as the best and animal welfare as the worst items in response to the 1st question.

A dataset suitable for analyzing responses to BWS questions is created using the `bws.dataset()` function. The arguments for `bws.dataset()` are determined according to the respondent dataset and choice sets as follows: `mfa` is assigned to argument `respondent.dataset`, `1` is assigned to argument `response.type` because row number format is applied to the respondent dataset, `sets.mfa` is assigned to argument `choice.sets`, and `1` is assigned to argument `design.type` because the choice sets are generated by a two-level OMED approach. Then, the dataset to be used in the analysis is generated using `bws.dataset()` and stored in the data frame `data.mfa`:

```
R> data.mfa <- bws.dataset(
                   respondent.dataset = mfa,
                   response.type = 1, # row number format
                   choice.sets = sets.mfa,
                   design.type = 1)   # OMED
R> head(data.mfa)
  Q PAIR ID RES.B RES.W BEST WORST ITEM1 ITEM2 ITEM3 ITEM4
1 1    1  1     9     8    2     5     0     1     0     0
2 1    2  1     9     8    2     7     0     1     0     0
3 1    3  1     9     8    2     8     0     1     0     0
4 1    4  1     9     8    2     9     0     1     0     0
5 1    5  1     9     8    5     2     0    -1     0     0
6 1    6  1     9     8    5     7     0     0     0     0
  ITEM5 ITEM6 ITEM7 ITEM8 ITEM9   RES STR
1    -1     0     0     0     0 FALSE 101
2     0     0    -1     0     0 FALSE 101
3     0     0     0    -1     0 FALSE 101
4     0     0     0     0    -1 FALSE 101
5     1     0     0     0     0 FALSE 101
6     1     0    -1     0     0 FALSE 101
```

As illustrated and explained in Section 4.3.4, the data frame contains various variables. Here, since the number of items is nine, there are nine state variables, `ITEM1` to `ITEM9`.

Based on the dataset, count-based scores are calculated using function `bws.count()` with argument `data = data.mfa`:

```
R> count.mfa <- bws.count(data = data.mfa)
R> count.mfa
Number of respondents = 100

Summary of disaggregated best-worst scores:
      meanB meanW meanBW mean.stdBW stdev.stdBW
ITEM1  1.90  0.71   1.19     0.1983       0.251
ITEM2  3.07  0.23   2.84     0.4733       0.249
ITEM3  0.56  2.17  -1.61    -0.2683       0.269
ITEM4  1.24  0.83   0.41     0.0683       0.232
```

```
ITEM5  0.22  3.26  -3.04     -0.5067          0.218
ITEM6  1.41  1.19   0.22      0.0367          0.229
ITEM7  0.55  1.94  -1.39     -0.2317          0.255
ITEM8  0.88  1.20  -0.32     -0.0533          0.198
ITEM9  2.17  0.47   1.70      0.2833          0.220

Aggregated best-worst scores:
        B   W   BW   stdBW sqrtBW std.sqrtBW
ITEM1 190  71  119  0.1983  1.636     0.4478
ITEM2 307  23  284  0.4733  3.653     1.0000
ITEM3  56 217 -161 -0.2683  0.508     0.1390
ITEM4 124  83   41  0.0683  1.222     0.3346
ITEM5  22 326 -304 -0.5067  0.260     0.0711
ITEM6 141 119   22  0.0367  1.089     0.2979
ITEM7  55 194 -139 -0.2317  0.532     0.1457
ITEM8  88 120  -32 -0.0533  0.856     0.2344
ITEM9 217  47  170  0.2833  2.149     0.5881
```

The means of the standardized BW scores are illustrated in Figure 4.7, which is drawn using the `barplot()` function:

```
R> par(mar = c(5, 9, 2, 2)) # change margins of plot
R> o <- order(count.mfa$aggregate$stdBW) # permutation of stdBW
R> barplot(height = count.mfa$aggregate$stdBW[o],
           xlim = c(-1, 1),            # limits of x-axis
           names.arg = items.mfa[o], # names of each bar
           horiz = TRUE,    # bars are drawn horizontally
           las = 2)         # labels are perpendicular to axis
```

Here, for ease of comparison of the scores for the nine roles, state variables (i.e., `ITEM1`, `ITEM2`, ..., `ITEM9`) are labeled according to the names of the roles using argument `names.arg`, while the roles are permuted according to the scores using the results of the `order()` function. On average, biodiversity, cultural heritage, landscape, land conservation, and rural viability have positive standardized BW scores, meaning that these roles were more frequently selected as the most important than the least important. According to these scores, the most important role of agricultural multifunctionality is, on average, biodiversity, the second-most important is cultural heritage, and the least important is flood control.

Next, the importance of each role is estimated by means of a CL model using `clogit()`. Assume that the coefficient of `ITEM6` (rural viability) is normalized to zero and that the analysis is conducted based on the dataset `data.mfa`:

```
R> fr.mfa <- RES ~ ITEM1 + ITEM2 + # exclude ITEM6 to
                   ITEM3 + ITEM4 + #  normalize its
                   ITEM5 + ITEM7 + #  coefficient to 0
                   ITEM8 + ITEM9 +  strata(STR)
```

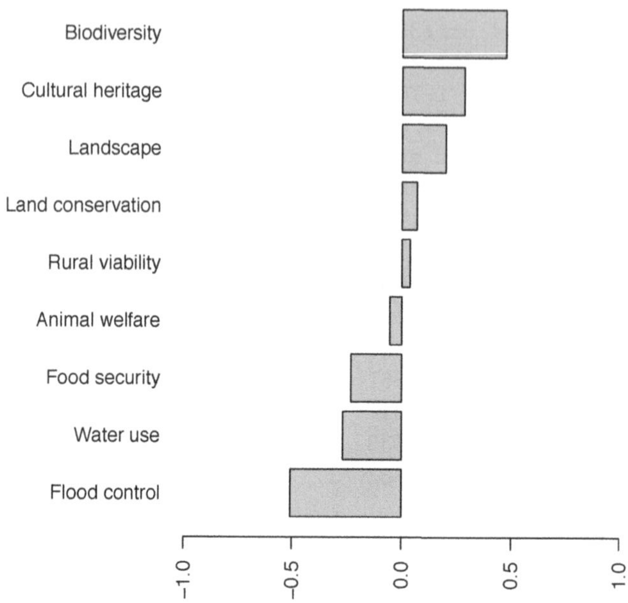

Figure 4.7 *Standardized BW scores for the multifunctionality of agriculture.*

```
R> clg.mfa <- clogit(formula = fr.mfa, data = data.mfa)
R> clg.mfa
Call:
clogit(formula = fr.mfa, data = data.mfa)


        coef exp(coef) se(coef)      z       p
ITEM1  0.378     1.459   0.0920   4.10 4.1e-05
ITEM2  1.106     3.021   0.0944  11.71 0.0e+00
ITEM3 -0.762     0.467   0.0921  -8.27 1.1e-16
ITEM4  0.116     1.122   0.0934   1.24 2.2e-01
ITEM5 -1.346     0.260   0.0960 -14.01 0.0e+00
ITEM7 -0.628     0.534   0.0923  -6.80 1.1e-11
ITEM8 -0.267     0.766   0.0926  -2.88 3.9e-03
ITEM9  0.620     1.858   0.0908   6.82 9.0e-12

Likelihood ratio test=983  on 8 df, p=0  n= 21600, number of
events= 1200
```

The `coef` column shows that the estimated coefficients of `ITEM1`, `ITEM2`, `ITEM4`, and `ITEM9` are positive, while those of `ITEM3`, `ITEM5`, `ITEM7`, and `ITEM8` are negative. However, the `p` column indicates that all variables other than `ITEM4` differ significantly from zero at the 1% level. Thus, the importance of land conservation (`ITEM4`) is not statistically different from that of "rural viability" (`ITEM6`) because the coefficient of `ITEM6` has been normalized to zero.

Let us compare the estimated coefficients with the aggregated standardized BW score (`stdBW`) contained in component `aggregate` of object `count.mfa`. The estimated coefficients are stored in the vector `coefficients`, which is contained in the output of `clogit()` (in this case, object `clg.mfa`):

```
R> clg.mfa$coefficients
  ITEM1   ITEM2   ITEM3   ITEM4   ITEM5   ITEM7   ITEM8
 0.3776  1.1056 -0.7620  0.1155 -1.3458 -0.6277 -0.2672
  ITEM9
 0.6196
```

Here, a coefficient of `0` that corresponds to the normalized item (in this case, `ITEM6`) has to be inserted into the appropriate element in the vector:

```
R> clgcoef.mfa <- c(clg.mfa$coef[1:5],
                    ITEM6 = 0,
                    clg.mfa$coef[6:8])
R> clgcoef.mfa
  ITEM1   ITEM2   ITEM3   ITEM4   ITEM5   ITEM6   ITEM7
 0.3776  1.1056 -0.7620  0.1155 -1.3458  0.0000 -0.6277
  ITEM8   ITEM9
-0.2672  0.6196
```

Then, `clogcoef.mfa` and `stdBW` are combined into the matrix `comp.mfa`:

```
R> comp.mfa <- cbind(clogit = clgcoef.mfa,
                     stdBW = count.mfa$aggregate$stdBW)
```

After the row names of `comp.mfa` have been changed according to the names of the nine roles (`items.names`), the comparison is made:

```
R> rownames(comp.mfa) <- items.mfa
R> comp.mfa
                     clogit    stdBW
Landscape            0.3776  0.19833
Biodiversity         1.1056  0.47333
Water use           -0.7620 -0.26833
Land conservation    0.1155  0.06833
Flood control       -1.3458 -0.50667
Rural viability      0.0000  0.03667
Food security       -0.6277 -0.23167
Animal welfare      -0.2672 -0.05333
Cultural heritage    0.6196  0.28333
```

The relationship between the two vectors `clogit` and `stdBW` stored in `comp.mfa` is shown in Figure 4.8, which is drawn using the `plot()` function:

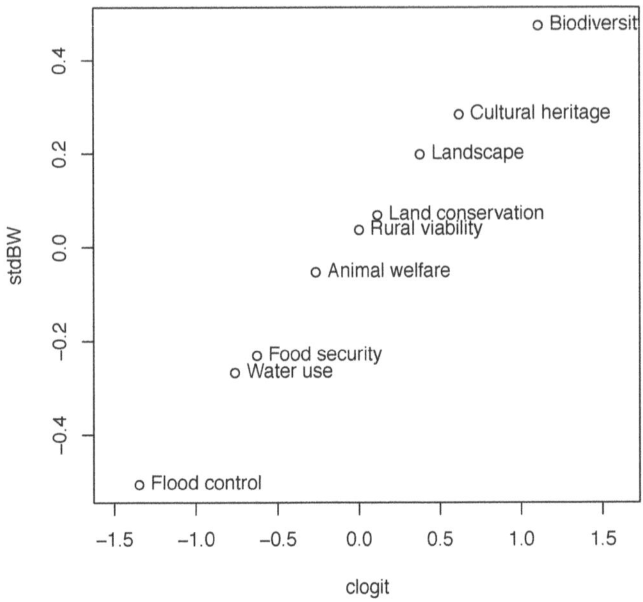

Figure 4.8 *Relationship between the conditional logit estimates and standardized BW scores.*

```
R> plot(comp.mfa, xlim = c(-1.5, 1.6))
```

Then, the `text()` function is used to display the names of roles to the right of the points in the graph:

```
R> text(comp.mfa, pos = 4,  # show names of roles to
        labels = items.mfa) #  the right of the points
```

As expected, the correlation between the two vectors is extremely high:

```
R> cor(comp.mfa)
       clogit  stdBW
clogit 1.0000 0.9994
stdBW  0.9994 1.0000
```

Finally, the shares of preference for each role are calculated. First, the numerator and denominator of the share of preference (see Eq. (4.8)) are calculated:

```
R> exp.clgcoef.mfa <- exp(clgcoef.mfa) # numerator
R> den <- sum(exp.clgcoef.mfa)         # denominator
```

Then, the shares of preference are calculated and converted into a matrix with row and column names for ease of use:

```
R> sp <- matrix(exp.clgcoef.mfa / den, # shares of preference
                nrow = 9, ncol = 1,    # 9 rows x 1 column
                dimnames = list(items.mfa,    # set row and
                                c("shares"))) #  column names
```

The resultant shares with their sum are shown using the `addmargins()` function:

```
R> addmargins(sp, 1) # show shares of preferences with their sum
                      shares
Landscape            0.13911
Biodiversity         0.28807
Water use            0.04451
Land conservation    0.10704
Flood control        0.02482
Rural viability      0.09536
Food security        0.05090
Animal welfare       0.07300
Cultural heritage    0.17719
Sum                  1.00000
```

As explained in Section 4.2.2, the share of preference is an easy-to-interpret scale. For example, the most important role of biodiversity (0.28807) and the second-most important role of cultural heritage (0.17719) are approximately twelve and seven times more important than the least important role of flood control (0.02482), respectively.

4.4.2 BWS based on a BIBD

The example of BWS using a BIBD is a hypothetical study that explores consumers' preferences for fruits. In this example, consumers were asked to evaluate seven fruits:

- Apple
- Orange
- Grapes
- Banana
- Peach
- Melon
- Pear

As stated in the literature (see e.g., Cochran and Cox, 1957), some BIBDs can handle seven items. Thus, this example uses one with seven items, four items per question, and seven questions. The BIBD is stored in the matrix `sets.fruit`:

```
R> sets.fruit <- cbind( # 7 rows x 4 columns
     c(1, 2, 2, 1, 1, 3, 1), # 1st column
```

Q1. Please select your most preferred fruit and your least preferred fruit.

Most	Fruit	Least
[ ]	Apple	[ ]
[ ]	Strawberry	[ ]
[ ]	Banana	[ ]
[ ]	Pear	[ ]

Figure 4.9 *Example BWS question for evaluating fruits.*

```
    c(4, 3, 4, 2, 3, 5, 2), # 2nd column
    c(6, 4, 5, 5, 4, 6, 3), # 3rd column
    c(7, 6, 7, 6, 5, 7, 7)) # 4th column
R> sets.fruit
     [,1] [,2] [,3] [,4]
[1,]    1    4    6    7
[2,]    2    3    4    6
[3,]    2    4    5    7
[4,]    1    2    5    6
[5,]    1    3    4    5
[6,]    3    5    6    7
[7,]    1    2    3    7
```

This design is the same as that in Table 4.2. See Section 4.3.2 for the code used to generate it.

The names of the seven fruits are stored in the vector `items.fruit`:

```
R> items.fruit <- c(
    "Apple",  # corresponds to value "1" in sets.fruit
    "Orange", # corresponds to value "2" in sets.fruit
    "Grapes", # corresponds to value "3" in sets.fruit
    "Banana", # corresponds to value "4" in sets.fruit
    "Peach",  # corresponds to value "5" in sets.fruit
    "Melon",  # corresponds to value "6" in sets.fruit
    "Pear")   # corresponds to value "7" in sets.fruit
```

Note that the order of names in the vector must correspond to the ascending order of level values in the BIBD.

According to these settings, BWS questions (Figure 4.9) are created using the `bws.questionnaire()` function:

```
R> bws.questionnaire(choice.sets = sets.fruit,
                     design.type = 2,  # BIBD
                     item.names  = items.fruit)
```

```
Q1
 Best Items  Worst
 [ ]  Apple  [ ]
 [ ]  Banana [ ]
 [ ]  Melon  [ ]
 [ ]  Pear   [ ]

...

Q7
 Best Items  Worst
 [ ]  Apple  [ ]
 [ ]  Orange [ ]
 [ ]  Grapes [ ]
 [ ]  Pear   [ ]
```

Each of the 100 consumers was asked to respond to all seven BWS questions by selecting the most preferred and the least preferred fruits in each question. Their responses were organized according to row number format and stored in the **support.BWS** package as the dataset `fruit`:

```
R> data(fruit)
R> head(fruit)
  ID B1 W1 B2 W2 B3 W3 B4 W4 B5 W5 B6 W6 B7 W7
1  1  1  2  1  4  2  3  1  4  4  1  2  3  2  4
2  2  1  3  4  3  4  3  3  4  1  4  4  3  1  2
3  3  2  3  1  4  2  3  2  4  3  2  4  1  2  3
4  4  1  2  2  4  3  4  1  4  1  3  2  3  1  3
5  5  4  2  1  3  1  2  2  3  1  3  4  2  2  1
6  6  1  2  1  2  3  2  3  4  4  2  4  1  2  3
```

For example, variables `B1` and `W1` in the first row of `fruit` contain `1` and `2`, respectively. This means that the respondent with `ID = 1` selected apple as the most preferred and banana as the least preferred fruits in the 1st question.

Based on the above settings, a dataset suitable for analysis is constructed using the `bws.dataset()` function and assigned to the object `data.fruit`:

```
R> data.fruit <- bws.dataset(
                   respondent.dataset = fruit,
                   response.type = 1,  # row number format
                   choice.sets = sets.fruit,
                   design.type = 2,    # BIBD
                   item.names = items.fruit)
R> head(data.fruit)
  Q PAIR ID RES.B RES.W BEST WORST Apple Orange Grapes
1 1    1  1     1     4    1     4     1      0      0
2 1    2  1     1     4    1     6     1      0      0
3 1    3  1     1     4    1     7     1      0      0
```

```
4 1    4 1     1     4    4     1    -1      0      0
5 1    5 1     1     4    4     6     0      0      0
6 1    6 1     1     4    4     7     0      0      0
  Banana Peach Melon Pear   RES STR
1     -1     0     0    0  TRUE 101
2      0     0    -1    0 FALSE 101
3      0     0     0   -1 FALSE 101
4      1     0     0    0 FALSE 101
5      1     0    -1    0 FALSE 101
6      1     0     0   -1 FALSE 101
```

Here, since the `bws.function()` was executed with argument `item.names = items.fruit`, the state variables are described using the names of the fruits (i.e., `Apple`, `Orange`, ..., `Pear`), instead of using default variables (i.e., `ITEM1`, `ITEM2`, ..., `ITEM7`).

Consumers' preferences for the seven fruits were estimated using a CL model. After defining the model `formula` based on the assumption that the coefficient of `Pear` was normalized to zero, the `clogit()` function was executed:

```
R> fr.fruit <- RES ~ Apple + Orange +  # exclude Pear to
                     Grapes + Banana + #  normalize its
                     Peach + Melon +   #  coefficient to 0
                     strata(STR)
R> clg.fruit <- clogit(formula = fr.fruit, data = data.fruit)
R> clg.fruit
Call:
clogit(formula = fr.fruit, data = data.fruit)


          coef exp(coef) se(coef)      z       p
Apple   1.1238     3.076   0.1060 10.604 0.0e+00
Orange  0.6103     1.841   0.1014  6.020 1.7e-09
Grapes  0.0998     1.105   0.0986  1.013 3.1e-01
Banana -0.0247     0.976   0.1015 -0.244 8.1e-01
Peach  -0.0195     0.981   0.0997 -0.196 8.4e-01
Melon  -0.9626     0.382   0.1058 -9.094 0.0e+00

Likelihood ratio test=447  on 6 df, p=0  n= 8400, number of
events= 700
```

Since the coefficient of `Pear` was normalized to zero, the preferences for other fruits are relative to pear. On average, the most preferred fruit is apple and the second-most preferred is orange; the least preferred is melon; while preferences for grapes, banana, and peach are not much different from that for pear because the coefficients of `Grapes`, `Banana`, and `Peach` do not differ significantly from zero.

Graphs are useful for showing heterogeneity in respondents' preferences toward each item (e.g., Mueller and Rungie, 2009). To draw these graphs, we first execute the counting approach:

```
R> count.fruit <- bws.count(data = data.fruit)
R> count.fruit
Number of respondents = 100

Summary of disaggregated best-worst scores:
       meanB meanW meanBW mean.stdBW stdev.stdBW
Apple   2.13  0.15   1.98     0.4950       0.286
Orange  1.49  0.51   0.98     0.2450       0.386
Grapes  1.08  1.12  -0.04    -0.0100       0.492
Banana  0.72  1.02  -0.30    -0.0750       0.374
Peach   0.67  0.92  -0.25    -0.0625       0.326
Melon   0.16  2.28  -2.12    -0.5300       0.271
Pear    0.75  1.00  -0.25    -0.0625       0.308

Aggregated best-worst scores:
         B   W   BW   stdBW sqrtBW std.sqrtBW
Apple  213  15  198  0.4950  3.768     1.0000
Orange 149  51   98  0.2450  1.709     0.4536
Grapes 108 112   -4 -0.0100  0.982     0.2606
Banana  72 102  -30 -0.0750  0.840     0.2230
Peach   67  92  -25 -0.0625  0.853     0.2265
Melon   16 228 -212 -0.5300  0.265     0.0703
Pear    75 100  -25 -0.0625  0.866     0.2298
```

Then, the code for the required graphs is executed based on the BW scores. As an example, let us draw Figure 4.10, showing the distributions of BW scores by item, based on disaggregated BW scores. The number of respondents for each score, which takes an integer value ranging from -4 to 4, by fruit, is calculated and stored in the list `score.tables`:

```
R> BW <- data.frame(count.fruit$disaggregate$BW)
R> BW <- lapply(BW, factor,           # scores are transformed
                levels = c(-4:4))     #  into a factor
R> score.tables <- lapply(BW, table) # table for each item
```

After storing the maximum number of respondents for all score values, which is used to define the upper limit of the y-axis in the following plots, the distribution of BW scores by item are plotted using the `barplot()` function:

```
R> ylimit <- max(unlist(score.tables)) # max count among items
R> par(mfrow = c(4,2)) # create 4 x 2 figure regions
R> for(i in 1:7){ # for each item
     barplot(height = score.tables[[i]],
             main = items.fruit[i], # title
```

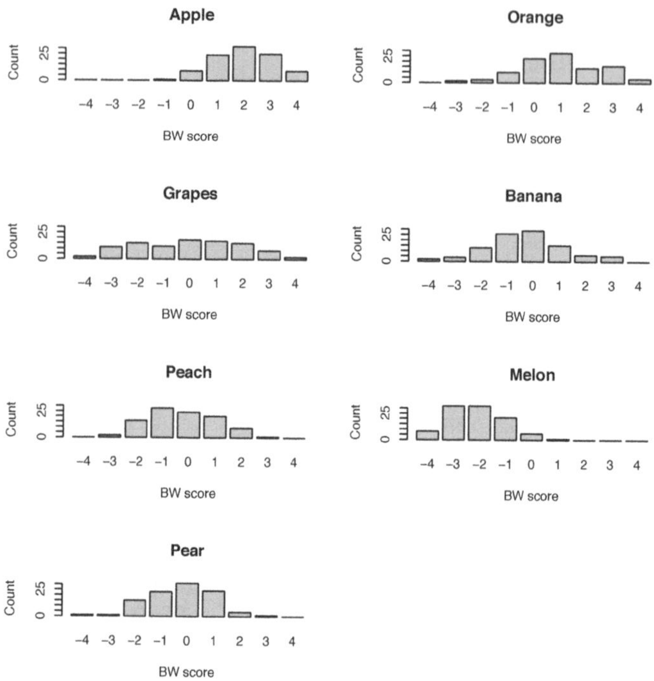

Figure 4.10 *Distributions of simple BW scores by items.*

```
            xlab = c("BW score"),  # label for x-axis
            ylab = c("Count"),     # label for y-axis
            ylim = c(0, ylimit))   # limits for y-axis
}
```

Although the BW scores range between −4 and +4 in this example, the extreme values are not that frequently shown in each item.

As another example, the following code displays Figure 4.11, showing the relationship between the mean BW score and its standard deviation, by item, based on the mean and standard deviation of disaggregated BW scores:

```
R> Mean.BW <- colMeans(count.fruit$disaggregate$BW)
R> Standard.Deviation.BW <- apply(count.fruit$disaggregate$BW,
                                   2, sd)
R> plot(x = Mean.BW,
        y = Standard.Deviation.BW,
        xlim = c(-3, 3),
        ylim = c(1, 2))
R> text(x = Mean.BW,
        y = Standard.Deviation.BW,
```

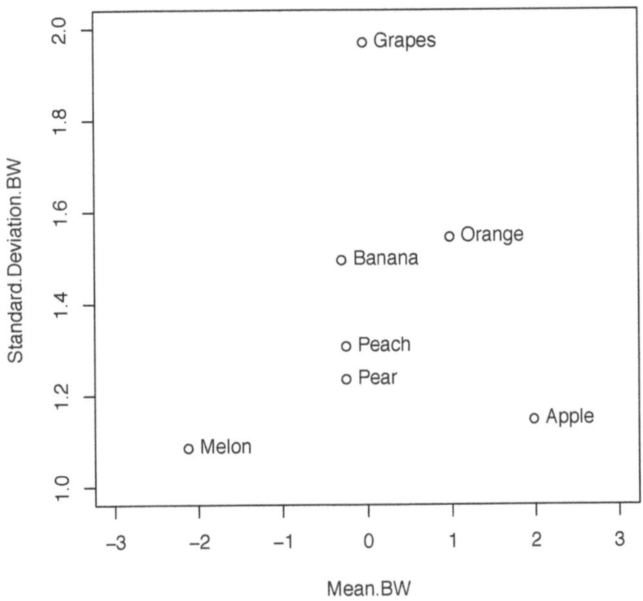

Figure 4.11 *Relationship between the mean BW score and its standard deviation.*

```
       pos = 4,
       labels = items.fruit)
```

The means of the preferences for pear, peach, banana, and grapes are similar to each other, while heterogeneity of the preference for grapes is relatively high. Apple has a relatively high mean with relatively low standard deviation of the score, which means that the majority prefer apples; whereas melon has a relatively low mean with low standard deviation of the score, meaning that the majority have a lesser preference for melon.

Finally, we divided the respondents into two groups and compared the preference each group has for the fruits. The respondents were classified into two groups using k-means clustering, which divides respondents into k clusters (groups) according to their within group sum of squares. The method can be executed using the `kmeans()` function in the `stats` package.[10] We set the disaggregated BW scores (`BW`) as the data for clustering and the number of clusters to 2, and then executed the function:

[10]See the application help for details of the `kmeans()` function.

```
R> set.seed(123)
R> kmeans.fruit <- kmeans(x = count.fruit$disaggregate$BW,
                          centers = 2)
```

The resultant means of the BW scores per item in each cluster are stored in the vector `centers` as the output of `kmeans.fruit`:

```
R> kmeans.fruit$centers
  Apple Orange Grapes  Banana    Peach  Melon     Pear
1 1.756 1.6444 -1.822  0.5556  0.08889 -2.244  0.02222
2 2.164 0.4364  1.418 -1.0000 -0.52727 -2.018 -0.47273
```

The mean BW scores are shown in Figure 4.12, which is drawn using the `barplot()` function:

```
R> barplot(height = kmeans.fruit$centers,
           names.arg = items.fruit,
           beside = TRUE, # draw juxtaposed bar
           legend = c("cluster.1", "cluster.2"),
           horiz = TRUE,
           las = 2,
           xlab = "BW score",
           xlim = c(-4, 4))
```

There are relatively larger differences in the BW scores between the two clusters with respect to grapes, banana, and orange, whereas there are small differences with respect to melon and apple.

4.5 Concluding remarks

This chapter explained how R functions can be used for basic object case BWS. Since BWS elicits individuals' relative importance of many items, wider empirical research fields should focus on the application of BWS. As introduced in Section 4.2.1, the ability of BWS could be strengthened by combining it with other methods such as factor analysis or cluster analysis. Readers who are interested in these combinations should refer to Appendix A, where various R packages related to these methods are briefly introduced.

4.A Appendix: Profile case BWS and multiprofile case BWS

In this chapter's appendix we briefly explain profile case BWS and multiprofile case BWS.

4.A.1 Profile case BWS

Profile case BWS is also known as Case 2 BWS, attribute-level BW choice experiments, attribute case BWS, or BW attribute scaling. A choice set, called a "profile" in DCE or conjoint analysis, is provided to respondents, who are asked to select the best and the worst attribute levels in the choice set. Whereas

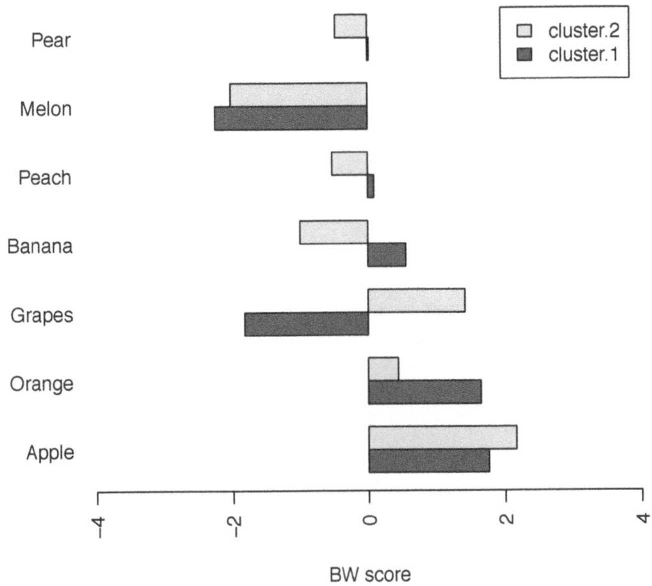

Figure 4.12 *Mean BW score per item in each cluster.*

object case BWS varies the combinations of items presented to respondents according to the questions, profile case BWS changes the combinations of attribute levels presented to them according to the questions (i.e., the attributes are fixed in all questions).

Figure 4.13 provides an example of profile case BWS designed to determine which characteristics consumers consider valuable when purchasing laptop computers. Laptop computers have four attributes, and each attribute has four attribute levels. In a question, a laptop computer is defined by combining the four attribute levels and is presented to respondents in a profile format. Then, respondents are asked to select the best and the worst attribute levels from the four presented. The profile format is the same as that used in binary DCEs with an opt-out option (see Chapter 3). However, whereas respondents are requested to either select or not select the "profile" shown to them in binary DCEs, they are requested to select the best and worst "attribute levels" in profile case BWS. Previous papers applying profile case BWS include Coast et al. (2006, 2008), Flynn et al. (2007, 2008), and Potoglou et al. (2011).

1) Setting attributes and attribute levels

Attribute	Level 1	Level 2	Level 3	Level 4
Body size	Under B5	B5	A4	Over A4
Memory	1 GB	2 GB	4 GB	8 GB
CPU	Brand A / 2 cores	Brand A / 4 cores	Brand B / 2 cores	Brand B / 4 cores
Storage	500 GB HD	1 TB HD	64 GB SSD	128 GB SSD

2) Constructing profiles by combining attribute levels using the design of experiments.

3) Asking respondents to select the best and worst attribute levels in each question.

Q1. Please select the characteristic that is the best for you and the characteristic that is the worst for you when purchasing laptop computers.

Best	Characteristic	Worst
[ ]	A4 size	[ ]
[ ]	8 GB memory	[ ]
[ ]	Brand A / 2 cores CPU	[✓]
[✓]	128 GB SSD	[ ]

Figure 4.13 *An example of profile case BWS together with a sample question.*

4.A.2 Multiprofile case BWS

Multiprofile case BWS is also known as Case 3 BWS or BWDCEs. In each question, three or more profiles are presented to a respondent, who is then asked to select the best and the worst profiles for each question (Figure 4.14).

The BWS format is the same as that used in DCEs. However, DCEs usually request respondents to select one profile or an alternative (the most preferred one), whereas multiprofile case BWS requests them to select the most preferred and the least preferred profiles (respondents may also be requested to rank the other profiles when four or more profiles are presented in each ques-

	Laptop PC 1	Laptop PC 2	Laptop PC 3
Body size	A4	A5	Over A4
Memory	1 GB	4 GB	2 GB
CPU	Brand A / 4 cores	Brand A / 2 cores	Brand A / 4 cores
Storage	500 GB HD	1 TB HD	64 GB SSD
Best	[ ]	[ ]	[✓]
Worst	[ ]	[✓]	[ ]

Figure 4.14 *An example question in multiprofile case BWS.*

tion). Theoretical foundations have only recently been developed (Marley and Pihlens, 2012), and thus there is limited literature addressing applications of this type of BWS (e.g., Louviere et al., 2008; Marley and Pihlens, 2012; Lancsar et al., 2013).

Chapter 5

Basic Operations in R

Abstract

Chapter 5 focuses on basic operations implemented in R. Since there are many good reference books at the introductory level, the explanations that follow are fairly limited. The topics covered include getting started with R, importing and exporting data, manipulating vectors and matrices, data and object types, implementing linear regression, drawing figures, and reading and loading source code.

5.1 Introduction

This chapter provides some information on how basic operations are carried out in R. Since there are many good reference books at the introductory level, the explanations that follow are fairly limited. If the user has any difficulty understanding the content of the subsequent sections, the first resource to consult is possibly Venables et al. (2014), particularly Appendix A. This manual as well as other useful documents are freely available from the website for R (`http://www.r-project.org/`).

5.2 Getting started with R

5.2.1 Obtaining and installing R

The base system for R is distributed in two different formats, namely as source code or precompiled binaries. The latter is available for different operating systems such as Windows (XP or later, 32- or 64-bit version), Mac OS X, and Linux (Debian or Ubuntu). The source code can also be used on these operating systems, but must be compiled before use. It is therefore sufficient for most users to download one of the precompiled binaries and install it on their personal computers (PCs).

All the distribution formats are downloadable from the Comprehensive R Archive Network (CRAN at `http://CRAN.R-project.org/`). The installation procedure depends on the operating system of the PC on which R is to be installed; please follow the instruction obtained from the CRAN.

5.2.2 Starting and quitting R

Once R has been installed on their PCs, users can enjoy using it. The following text is displayed on the R console immediately after starting R:

```
R version 3.0.1 (2013-05-16) -- "Good Sport"
Copyright (C) 2013 The R Foundation for Statistical Computing
Platform: x86_64-apple-darwin10.8.0 (64-bit)

R is free software and comes with ABSOLUTELY NO WARRANTY.
You are welcome to redistribute it under certain conditions.
Type 'license()' or 'licence()' for distribution details.

  Natural language support but running in an English locale

R is a collaborative project with many contributors.
Type 'contributors()' for more information and
'citation()' on how to cite R or R packages in publications.

Type 'demo()' for some demos, 'help()' for on-line help, or
'help.start()' for an HTML browser interface to help.
Type 'q()' to quit R.

>
```

The last row displays a > symbol. This is a prompt, showing that R is waiting for the user to execute a command or input R code.

To quit the current R session, type

```
R> q()
```

whereupon the user is asked whether to save the workspace: typing `y` (yes) or `n` (no), respectively, saves the workspace before quitting or just quits the R session. The same can be achieved by executing the following:

```
R> q(save="yes")  # saving the current workspace
```

and

```
R> q(save="no")   # not saving the workspace
```

When requesting R to save the workspace, all the objects and the command history are saved in a `.RData` file and a `.Rhistory` file, respectively, in the current working directory. These files are automatically loaded into the workspace when a new R session is started. If these files do not need to be loaded, they should be removed from the current working directory.

5.2.3 Using R as a calculator

The simplest use of R is as a calculator. The four arithmetic operations can be executed in R using the + (addition), - (subtraction), * (multiplication), and / (division) operators. For example, the following line of code returns the result of 2 multiplied by 3:

```
R> 2 * 3
[1] 6
```

If the input code extends over two or more successive lines, the character used as the prompt in the second and subsequent lines is the + symbol. The following two lines of code return the same result as the code given above:

```
R> 2 *
   3
[1] 6
```

However, the user should be careful when writing code extending over multiple lines that only the last line ends with a fragment of code that can be regarded as the end of the command. This is not always the case when code is written in an external code editor and "copy-pasted" into the R console.

This is illustrated by a simple example. Suppose the user wants to compute

$$2 + 3 + 4 + 5$$

in R. This can be achieved by the first command given below, but not by the second one. That is,

```
2 + 3 + 4 + 5
```

and

```
2 + 3
+ 4 + 5
```

return different results. The first piece of code calculates `2 + 3 + 4 + 5`, while the second executes two operations: the first line calculates `2 + 3` while the second calculates `4 + 5`. If this code has to be split over several lines, it should be written as follows:

```
2 + 3 +
4 + 5
```

Putting the plus operator (+) at the end of the first line tells R that the line continues to the next one. This is a simple example, so the user will most likely recognize the erroneous outcome immediately and be able to pinpoint the bug. However, it could be extremely difficult to identify such a bug if it occurred in the middle of a lengthy piece of code spread over multiple lines. A more serious problem could arise if this type of operation were embedded in a user-defined function and each piece of code was correct in terms of R syntax. An R function usually does not display intermediate results, and in the above case, it would not terminate in error because each piece was executed correctly.

5.2.4 *Changing appearance*

The `options()` function can be used to modify the prompt characters as well as other settings. For example, the following code sets the appearance format for R as used in this book:

```
R> options(prompt = "R> ", continue = "   ", width = 60,
           digits = 4, show.signif.stars = FALSE)
```

where argument `prompt` sets the prompt to "`R>` "; argument `continue` sets the prompt used in continuation lines to " "; argument `width` sets the maximum number of columns on a line to 60; argument `digits` sets the number of digits to be output to the display to 4; and argument `show.signif.stars` denotes that marks showing significance levels are not displayed in summary coefficient tables.

5.2.5 *Accessing help*

When requiring information on how to use a particular function, users can access help using `?` or `help()`. For example, help on `q()` is displayed by executing the following:

```
R> ?q
```

or

```
R> help(q)
```

5.3 Enhancing R

5.3.1 *Installing contributed add-on packages*

The base system for R contains the basic and recommended packages listed in Table 5.1. Besides these, R is enhanced with many contributed add-on packages. Briefly speaking, add-on packages are collections of R functions and datasets designed to implement particular statistical and/or graphic techniques. At the time of writing (end of July 2013), more than 4700 packages are available and the number is growing exponentially. There are dependent networks among the R packages, making it possible for one package to use functions and methods defined in another. In fact, this dependent mechanism is one of the key features in the R package system.

To output the names of the packages already installed in the R tree, run the following command

```
R> library()
```

and the list of installed packages will appear. If the user wants to use packages that do not exist in the system, they can be added through the R console by executing the `install.packages()` function. Suppose that the name of the desired package is **support.CEs**, then

```
R> install.packages("support.CEs", dependencies = TRUE)
```

Table 5.1 *R packages bundled with the base system*

Basic		Recommended	
base	**parallel**	**boot**	**Matrix**
compiler	**splines**	**class**	**mgcv**
datasets	**stats**	**cluster**	**nlme**
graphics	**stats4**	**codetools**	**nnet**
grDevices	**tcltk**	**foreign**	**rpart**
grid	**tools**	**KernSmooth**	**spatial**
methods	**utils**	**lattice**	

installs it into the R tree as well as the other packages on which **support.CEs** depends. Package installation can also be initiated from the menu. In the Windows version of R, for example, by selecting "Install package(s)..." from the "Packages" menu in the R GUI, the user will be prompted to select the desired packages from a list of available package names. Sometimes the user may be asked to set the CRAN mirror from a given list prior to selecting packages.

Packages must be loaded in each R session before they are used via the command:

```
R> library(support.CEs)
```

Dependent packages are loaded simultaneously.

Each CRAN package is assigned a URL for its source code, the PDF manual, vignette files if any, information on the author(s), and so on. The web page is accessible at `http://cran.r-project.org/package=yyy`, where `yyy` is the name of the package (which is case sensitive).

As the CRAN is mirrored worldwide,[1] the user can select the closest site from which to obtain information on particular packages. In terms of download speed, the RStudio[2] CRAN mirror (`http://cran.rstudio.com/`) is another alternative. It uses a cloud-based content delivery network, so files are retrieved from a local cloud server automatically. To use the RStudio mirror or others, replace `cran.r-project.org` with `cran.rstudio.com` or the like.

Besides CRAN and its mirrors, there are two more repositories for the distribution of R packages. One is R-Forge (`http://r-forge.r-project.org`) and the other is Bioconductor (`http://www.bioconductor.org`). R-Forge offers a platform for the development of R packages and R-related projects. Bioconductor provides R tools for bioinformatics analysis. Some of the add-on packages are available only from the R-Forge or Bioconductor sites. Installing

[1] A list of CRAN mirrors is available at `http://cran.r-project.org/mirrors.html`.

[2] Besides the CRAN mirror server, RStudio (`http://www.rstudio.com/`) offers a free and open source integrated development environment (IDE) with the same name for R. The RStudio IDE runs on Windows, Mac OS, and Linux as well as over the web. It has convenient functionalities such as syntax highlighting and direct execution of R code from the source editor.

packages from these repositories is not always possible using the GUI package installer. In fact, **DCchoice** explained in Chapter 2 is such an example. When installing packages from either of these two repositories, please read the installation procedures for the individual packages carefully.

5.3.2 Reading source code

It is sometimes useful to check the source code to see how an actual computation is carried out. To read the source code for a function, the function name without any parentheses is executed in R. For example, the following code displays the source code for the `lm()` function (note that the output has been partially modified):

```
R> lm
function (formula, data, subset, weights, na.action,
    method = "qr", model = TRUE, x = FALSE, y = FALSE,
    qr = TRUE, singular.ok = TRUE, contrasts = NULL, offset,
     ...)
{
    ret.x <- x
    ret.y <- y
...
```

More importantly, the source code is often interspersed with comments that may help the user understand the method more precisely. Because comments written in the source are removed when the package is installed, the only option is to check the source code directly. A package source in the CRAN is provided in compressed `tar.gz` format. A `tar.gz` file can be unpacked, for example, by using 7-zip (`http://www.7-zip.org/`) in Windows or the unarchiver (`http://unarchiver.c3.cx/unarchiver`) in Mac OS. The unarchived package folder (directory) contains several sub-folders. One of these is named `R`, and contains the R programs saved in files with the extension `.R`. Since these are standard text files, they can be opened with any text editor. RStudio can of course handle `.R` files as well.

5.3.3 Loading source code

Functions are sometimes distributed in text format. The `source()` function loads an external file containing R code and executes it. For example,

```
R> source("foo.R")
```

reads the file `foo.R` in the current working directory and executes the contents. Files can be placed in any location on the local computer or external device. If the file is located in a place other than the current working directory, the full pathname of the file should be given, for example, `"c:/foo/foo.R"` and the like.

The `source()` function can handle files stored on the Internet. A valid URL should be provided in this case.

5.4 Importing and exporting data

5.4.1 File formats

For Windows users, Microsoft Excel is the most widely used tool for handling and exchanging data. There are various packages that help R import data in Microsoft Excel formats (`.xls` and `.xlsx`). Details can be found, for example, in R Core Team (2014, Chapter 9). Other data formats such as SAS, SPSS, or Stata, to name but a few, can also be handled directly by R with the help of the **foreign** package.

However, as R Core Team (2014) pointed out, an ASCII or text file is the easiest data format for import into R. In a text format, each component of a data entry is usually separated either by a comma (,) or by white space (spaces or tabs). A text file using the comma separator is referred to as a comma-separated values (CSV) file. Most statistical software including those mentioned above are capable of exporting data as a CSV file. Therefore, the subsequent explanations focus on how to import data into R from CSV files.

There are also some other issues to consider, such as file encoding and the handling of non-ASCII characters (especially in Japanese/Chinese/Korean languages). If a user encounters such problems, he/she should in the first instance refer to R Core Team (2014). Alternatively, consulting local experts is also recommended.

5.4.2 Importing data from a CSV file

The simplest option for importing data from a CSV file is to use the `read.csv()` function, which is a convenient interface to the `read.table()` function. Both `read.csv()` and `read.table()` read not only CSV files but also standard ASCII data files into the current R session.

A typical command for importing data from a CSV file and assigning it to an object `sample.data` is given by:

```
R> sample.data <- read.csv(file = "sample_data.csv",
                          header = TRUE, na.strings = "NA")
```

where symbol `<-` denotes the assignment operator, by which the result of executing the code on the right side of the operator is assigned to the object on the left side (`sample.data` in the above example). The `read.csv()` function generates a data frame object from the CSV file, and thus `sample.data` is also a data frame.[3] The name of the CSV file is given in double quotes in the `file` argument. If the CSV file is saved in a directory other than the current working directory, this must be specified by adding either a relative path (`../terget_path/`) or an absolute one (`c:/terget_path/`) before the name of the CSV file.[4] Argument `header` specifies whether the first line of the file

[3]Data types including the data frame are explained in Section 5.6.

[4]The `getwd()` function is used to ascertain what the current working directory is, while `setwd()` is used to change the current working directory.

contains the names of variables. If `header = TRUE` (which is the default setting in `read.csv()`), the variable names are used as column names in `sample.data`. If the first data entry is stored in the first line of the CSV file, `header = FALSE` should be used. Otherwise, the components of the data entry in the first line are interpreted as column labels. Argument `na.strings` defines how missing components are labeled in the CSV file. If `na.strings = "NA"`, the character string `"NA"` in the CSV file is interpreted as a missing value. As long as it is specified correctly in `na.strings =`, missing components in the CSV file can be demarcated using any number or character string convenient to the user. Typical symbols for missing values are `-999` and `na`, among others. Once the CSV file has been loaded into a data frame object by `read.csv()`, missing components are represented by the special value `NA`.

When importing data from external files, it is advisable to see if the import was successful. The following commands

```
R> head(sample.data)
```

and

```
R> tail(sample.data)
```

display the first and the last few rows of the data, respectively. The command

```
R> dim(sample.data)
```

is also useful to check that the correct numbers of rows (observations) and columns (variables) have been imported.

5.4.3 *Exporting* R *objects*

The function used to export an R object as a CSV file is `write.csv()`. Usage of `write.csv()` is given by:

```
R> write.csv(sample.data, file = "new.sample.data.csv",
             row.names = FALSE)
```

This example code exports an R object `sample.data` as a CSV file saved as `new.sample.data.csv` in the current working directory. By default, `write.csv()` writes the row and column names with double quotes. This is not always convenient. Setting `quote = FALSE` and `row.names = FALSE` suppresses enclosing the row and column names with double quotes and exporting the row names, respectively.

The `write.csv()` function is a wrapper for the `write.table()` function. Other options can be examined by executing

```
R> ?write.csv
```

or

```
R> ?write.table
```

Both these commands display the same help file.

5.5 Manipulating vectors and matrices

5.5.1 Manipulating vectors

When manipulating data in R, a vector is the fundamental data type; a scalar is expressed as a vector with one element. The following code creates a vector `a`, containing the integer values from 1 to 5:

```
R> a <- 1:5
R> a
[1] 1 2 3 4 5
```

The length of a vector, which is the number of elements in the vector, is calculated using the `length()` function as follows:

```
R> length(a)
[1] 5
```

Two or more vectors of the same length can be added, subtracted, multiplied, and divided element-wise using the arithmetic operators `+`, `-`, `*`, and `/`, respectively. For example, addition of two vectors `a` and `b` is achieved by

```
R> b <- 6:10
R> a + b
[1]  7  9 11 13 15
```

The reader should be aware that even if the lengths of the vectors are not the same, addition and the other basic arithmetic operations are still executed. For example, adding vector `d` of length 2 to vector `a` of length 5

```
R> d <- 3:4
R> a + d
[1] 4 6 6 8 8
```

produces the following warning message:

```
Warning message:
In a + d : longer object length is not a multiple of shorter
object length
```

What actually happens in executing `a + d` is that elements `(3 4)` are recycled to create a vector of length 5, that is, `(3 4 3 4 3)`, which is then added to `a`.

If the longer object length is indeed a multiple of the shorter object length, no warning is produced. In the following code,

```
R> e <- 1:4
R> d + e
[1] 4 6 6 8
```

vector `d` is concatenated with itself to create a vector of length 4, which is the same as the length of `e`.

The same convention applies to element-wise operations with `+`, `-`, `*`, and `/`. Therefore, it is good practice to be aware of the length of objects.

Exponentiation of a vector using the `^` operator is done element-wise. Thus,

```
R> a^2
[1]  1  4  9 16 25
```

produces a vector with length equal to that of a where each element is the square of the corresponding element in a.

Vector objects can behave like matrix objects. The inner product of a single vector gives

```
R> a%*%a
     [,1]
[1,]   55
```

which is a 1×1 matrix. A similar result is obtained by

```
R> sum(a^2)
[1] 55
```

Although the resulting value is the same, the latter returns a vector of length one.

The outer product of a vector is computed by %o%.

```
R> a%o%a
     [,1] [,2] [,3] [,4] [,5]
[1,]    1    2    3    4    5
[2,]    2    4    6    8   10
[3,]    3    6    9   12   15
[4,]    4    8   12   16   20
[5,]    5   10   15   20   25
```

Operations on vectors, matrices, or data frames with NA elements should be handled with care. For example, if the sum() function is applied to a vector with an NA element, it returns NA unless na.rm = TRUE is explicitly stated:

```
R> a <- c(1, 2, NA, 4)
R> a
[1]  1  2 NA  4
R> sum(a)
[1] NA
R> sum(a, na.rm = TRUE)
[1] 7
```

By setting na.rm = TRUE, summation is taken over the vector a excluding NA and NaN (not a number) elements. The same convention is used for mean(), colSums(), and so on.[5]

[5]It is often the case that the user wants to remove observations with missing values in advance to create a new data frame. This is done using the na.omit() function.

5.5.2 *Manipulating matrices*

Two matrices with the same dimension can be added, subtracted, multiplied, and divided element-wise using the arithmetic operators `+`, `-`, `*`, and `/`, respectively.

Matrix multiplication is defined by `%*%` for two conforming matrices. The inverse and determinant of a square matrix are obtained by `solve()` and `det()`, respectively. A matrix object is created from an object, say `x`, by `matrix(x, nrow, ncol)`, where `nrow` and `ncol` specify the numbers of rows and columns, respectively. Object `x` can be an existing vector or matrix object. If the length of `x` is smaller than `nrow` $\times$ `ncol`, the first few elements of `x` are recycled to fill the gaps.

The following lines of code create two matrices, `A` and `B`, multiply `A` with `B`, compute the inverse of `A` with `B`, and return the determinant of `A`:

```
R> A <- matrix(rnorm(4), nrow = 2, ncol = 2)
R> A
        [,1]    [,2]
[1,]  0.7659 -0.1343
[2,] -1.6777 -0.6628
R> B <- matrix(rnorm(4), nrow = 2, ncol = 2)
R> B
        [,1]    [,2]
[1,] -0.8608  0.2049
[2,]  1.6504 -0.6039
R> A%*%B
        [,1]    [,2]
[1,] -0.8808 0.23799
[2,]  0.3502 0.05661
R> solve(A)%*%B
        [,1]   [,2]
[1,] -1.0808 0.2959
[2,]  0.2456 0.1622
R> det(A)
[1] -0.7329
```

The `matrix()` function by default inserts the elements of `x` column-wise, that is, the matrix is filled from element $(1, 1)$ to element $(nrow, 1)$, and then from $(1, 2)$ to $(nrow, 2)$, and so on. If the user wishes to fill the matrix in a row-wise fashion, argument `byrow = TRUE` should be specified as follows:

```
R> a <- 1:4
R> matrix(a, nrow = 2, ncol = 2)
     [,1] [,2]
[1,]    1    3
[2,]    2    4
```

```
R> matrix(a, nrow = 2, ncol = 2, byrow = TRUE)
     [,1] [,2]
[1,]    1    2
[2,]    3    4
```

It is also possible to create a matrix object by concatenating two or more vectors using `cbind()` or `rbind()`:

```
R> b <- 5:8
R> cbind(a, b)
     a b
[1,] 1 5
[2,] 2 6
[3,] 3 7
[4,] 4 8
R> rbind(a, b)
  [,1] [,2] [,3] [,4]
a    1    2    3    4
b    5    6    7    8
```

It should be noted that the number of rows in `cbind()` or the number of columns in `rbind()` is determined by the length of the longer vector. Once again, any shortage of elements is solved by recycling elements in the shorter vector.

5.5.3 Operations on indexes

Elements in a vector or matrix object can be accessed by specifying the index via the `[]` operator. Operations on indexes are indeed helpful to create subsets of vectors and matrices.

As an example, the first two elements of vector `a`, which contains integers from -3 to 5, can be extracted by `a[1:2]` as follows:

```
R> a <-c(-3:5)
R> a
[1] -3 -2 -1  0  1  2  3  4  5
R> a[1:2]
[1] -3 -2
```

Since the length of `a` is 9, the following operation produces exactly the same outcome.

```
R> a[-c(3:9)]
[1] -3 -2
```

This operation on the indexes excludes the third to ninth elements of `a`. Generally, assigning negative values in `[]` removes the corresponding elements. Moreover, the specified indexes need not be consecutive. That is,

```
R> a[c(1, 3, 5)]
[1] -3 -1  1
```

returns a vector whose elements are taken from the first, third, and fifth elements of a.

As another example, after creating matrix A from vector a:

```
R> A <- matrix(a, nrow = 3, ncol = 3)
R> A
     [,1] [,2] [,3]
[1,]   -3    0    3
[2,]   -2    1    4
[3,]   -1    2    5
```

the following lines of code show element $(1, 1)$, the second row, and the third column, respectively, of A:

```
R> A[1,1]
[1] -3
R> A[2, ]
[1] -2  1  4
R> A[, 3]
[1] 3 4 5
```

The last two commands above return vector objects. If the matrix property must be maintained, argument drop = FALSE is used as follows:

```
R> A[2, , drop = FALSE]
     [,1] [,2] [,3]
[1,]   -2    1    4
R> A[, 3, drop = FALSE]
     [,1]
[1,]    3
[2,]    4
[3,]    5
```

Sometimes the user needs to know the positions of elements satisfying certain conditions. For this purpose, the which() function is useful. For example, the following lines of code return the positions of elements in vector a that are exactly equal to or strictly greater than zero, respectively:

```
R> which(a == 0)
[1] 4
R> which(a > 0)
[1] 5 6 7 8 9
```

The results show that zero is stored in the fourth element of a, while values strictly greater than zero are stored in the fifth to ninth elements of a.

When the inspected object is a matrix, the function with argument arr.ind = TRUE returns the positions in the rows and columns as a vector or matrix object.

```
R> which(A > 0, arr.ind = TRUE)
     row col
[1,]   2   2
[2,]   3   2
[3,]   1   3
[4,]   2   3
[5,]   3   3
```

Executing this command shows that positive values in A are located at elements $(2, 2)$, $(3, 2)$, $(1, 3)$, $(2, 3)$, and $(3, 3)$.

Up to this point, no name has been assigned to the elements in vectors or the rows and columns of matrices. Elements in a vector can be named using the names() function. For example, the following code assigns small letters of the alphabet as the name of each element in vector a:

```
R> names(a) <- letters[1:9]
R> a
 a  b  c  d  e  f  g  h  i
-3 -2 -1  0  1  2  3  4  5
R> names(a)
[1] "a" "b" "c" "d" "e" "f" "g" "h" "i"
```

Once the elements in a vector have been named, they can also be accessed via their names as shown below:

```
R> a["b"]
 b
-2
R> a[c("c", "b", "a")]
 c  b  a
-1 -2 -3
```

The columns and rows of matrix A can be named using the functions colnames() and rownames(), respectively:

```
R> colnames(A) <- c("c1", "c2", "c3")
R> A
     c1 c2 c3
[1,] -3  0  3
[2,] -2  1  4
[3,] -1  2  5
R> rownames(A) <- c("r1", "r2", "r3")
R> A
   c1 c2 c3
r1 -3  0  3
```

```
r2 -2  1  4
r3 -1  2  5
```

The row and column names can be obtained by these functions:

```
R> colnames(A)
[1] "c1" "c2" "c3"
R> rownames(A)
[1] "r1" "r2" "r3"
```

As before, elements, rows, and columns in a matrix can be accessed using their names as follows:

```
R> A["r1", "c1"]
[1] -3
R> A["r2", ]
c1 c2 c3
-2  1  4
R> A[, "c3"]
r1 r2 r3
 3  4  5
```

It is also possible to change the name of a particular element in a vector and column or row in a matrix. For example, the following two lines of code change the name of the third element in the vector a to C, and the name of the first row in matrix A to R1, respectively:

```
R> names(a)[3] <- "C"
R> a
 a  b  C  d  e  f  g  h  i
-3 -2 -1  0  1  2  3  4  5
R> rownames(A)[1] <- "R1"
R> A
   c1 c2 c3
R1 -3  0  3
r2 -2  1  4
r3 -1  2  5
```

5.5.4 *Random number generation*

R provides several functions for generating random numbers based on various probability distributions such as the normal, Poisson, and uniform distributions, among others (see Venables et al. (2014) for a list of these functions). For example, the following R code generates five random numbers from the standard normal distribution:

```
R> rnorm(5)
[1]  1.77723 -1.28785 -0.08735  1.05878  0.68021
```

Functions such as `krCI()` and `bootCI()` in Chapter 2 for constructing confidence intervals by simulation internally use random number generators. These functions therefore generate different outcomes for repeated executions with the same object.

If the same results need to be replicated in future, the user should set the seed for the random number generator using function `set.seed()` before generating any random numbers:

```
R> set.seed(123)
R> rnorm(5)
[1] -0.56048 -0.23018  1.55871  0.07051  0.12929
R> set.seed(123)
R> rnorm(5)
[1] -0.56048 -0.23018  1.55871  0.07051  0.12929
```

It should be noted that the same value of the seed guarantees that the same sequence of random numbers is returned in different sessions on the same PC, although this is not necessarily the case on different PCs and for different versions of OSes/R itself.

5.6 Data and object types

5.6.1 Data types

Data that are used throughout the book are classified into the following categories:[6]

- *integer*: Integer values such as `1` or `2`.
- *numeric*: Real values such as `1.4142` or `3.1415`.
- *factor*: A special kind of integer value, that is, categorical values defined as a combination of a set of characters and integer values corresponding to the characters (see the following example for details).
- *character*: Character values that are enclosed by double quotation marks (e.g., `"123"` or `"xyz"`; note that the font used to depict the double quotation marks is that normally used for R code).
- *logical*: Logical values of `TRUE` and `FALSE`.

5.6.2 Object types

Items created by functions or even the functions themselves within R are known as *objects*. An object has its own types. The main types of objects are listed below:

- *vector*: An object with one or more elements, all of which must contain the same class of data.

[6]Explanations given above are roughly written from the viewpoint of R objects' features of "class." See R Core Team (2014) for a more detailed explanation.

- *array*: An object with two or more dimensions, all the elements of which must also contain the same class of data.
- *matrix*: A two-dimensional array.
- *list*: An object storing one or more different class objects; the data class of each component in the list can differ from that of the other components.
- *data frame*: A list with one or more variables, the length of which is the same.
- *function*: A function providing a convenient interface to a series of operations in R and returning another object. Operations are not limited to mathematical ones.

5.6.3 Examples

As an example, the following lines of code create object `A`, which is a list with five components: vector `a`, which contains sequential integers from 1 to 12; vector `b`, which contains twelve normally distributed random numbers; 3×4 matrix `c` with twelve logical values (`TRUE` or `FALSE`); vector `d` with six elements containing one of three characters, namely, `"X"`, `"Y"`, and `"Z"`; and vector `e` containing factor values with three levels, namely, `X`, `Y`, and `Z`.

```
R> a <- c(1:12)
R> b <- rnorm(12)
R> c <- matrix(b > 0.5, nrow = 3, ncol = 4)
R> d <- c("X", "Z", "Y", "X", "Z", "X")
R> e <- factor(d)
R> A <- list(a = a, b = b, c = c, d = d , e = e)
```

```
R> str(A)
List of 5
 $ a: int [1:12] 1 2 3 4 5 6 7 8 9 10 ...
 $ b: num [1:12] 1.715 0.461 -1.265 -0.687 -0.446 ...
 $ c: logi [1:3, 1:4] TRUE FALSE FALSE FALSE FALSE TRUE ...
 $ d: chr [1:6] "X" "Z" "Y" "X" ...
 $ e: Factor w/ 3 levels "X","Y","Z": 1 3 2 1 3 1
R> A
$a
 [1]  1  2  3  4  5  6  7  8  9 10 11 12


$b
 [1]  1.7151  0.4609 -1.2651 -0.6869 -0.4457  1.2241  0.3598
 [8]  0.4008  0.1107 -0.5558  1.7869  0.4979


$c
      [,1]  [,2]  [,3]  [,4]
[1,]  TRUE FALSE FALSE FALSE
```

```
[2,] FALSE FALSE FALSE  TRUE
[3,] FALSE  TRUE FALSE FALSE

$d
[1] "X" "Z" "Y" "X" "Z" "X"

$e
[1] X Z Y X Z X
Levels: X Y Z
```

Each component in a list object can be accessed using the [] or $ operators. Therefore, the following three lines of code return exactly the same output, that is, the first component in list object A:

```
R> A[1]
$a
 [1]  1  2  3  4  5  6  7  8  9 10 11 12
R> A["a"]
$a
 [1]  1  2  3  4  5  6  7  8  9 10 11 12
R> A$a
 [1]  1  2  3  4  5  6  7  8  9 10 11 12
```

As another example, the following code creates object B, which is a data frame containing vectors d and e:

```
R> B <- data.frame(d = d, e = e)
R> B
  d e
1 X X
2 Z Z
3 Y Y
4 X X
5 Z Z
6 X X
```

Each column in a data frame can be accessed using [] and $ operators. Each line of the code given below displays the same output, that is, the second column in B:

```
R> B[2]
  e
1 X
2 Z
3 Y
4 X
5 Z
6 X
```

```
R> B["e"]
  e
1 X
2 Z
3 Y
4 X
5 Z
6 X
R> B$e
[1] X Z Y X Z X
Levels: X Y Z
```

5.7 Implementing linear regression

The basic usage of R functions for statistical models is explained using an example based on the `lm()` function in package **stats**, which estimates linear models.

5.7.1 *Conducting the analysis*

R functions for statistical models generally have an argument `formula`, which is used to specify the structure of the model to be estimated based on symbolic expression. For example, the basic usage of the `lm()` function is given as

```
R> lm(formula, data)
```

where argument `data` is set as a data frame containing variables used in the model expression specified by `formula`. Note that other arguments have been omitted for simplicity.

Assume that we would like to estimate the linear model

$$y = \beta_0 + \beta_1 x1 + \beta_2 x2, \tag{5.1}$$

where y is a dependent variable; $x1$ and $x2$ are independent variables; and β_0, β_1, and β_2 are the coefficients to be estimated. The model expression corresponding to this model is:

```
R> y ~ x1 + x2
```

The left side of the tilde (~) denotes the dependent variable, while the right side denotes the independent variables. Where there are two or more independent variables, each independent variable is connected by a `+`. The `+` symbol should not be interpreted as addition of `x1` and `x2`. It simply means that the model includes `x1` and `x2` as independent variables. A constant term in the model does not need to be defined explicitly in the formula; it is automatically inserted when executing the `lm()` function. Further details of the syntax of formula are summarized in Venables et al. (2014).

Assume that a dataset for this model contains ten observations and is constructed as

```
R> lmdata <- data.frame(
        y  = c(11, 13, 32, 30, 27, 33, 40, 43, 24, 25),
        x1 = c(3, 4, 6, 9, 2, 9, 9, 7, 6, 1),
        x2 = c(4, 4, 14, 8, 15, 10, 14, 20, 8, 16))
```

Then, the following code implements the analysis based on the model and dataset:

```
R> lmout <- lm(y ~ x1 + x2, data = lmdata)
```

where the estimated model is stored in an object `lmout`.

5.7.2 Displaying and summarizing output

A class is generally assigned to the output from R functions. For example, the output from `lm()` is of the class "`lm`":

```
R> class(lmout)
[1] "lm"
```

R has generic functions that behave according to the type of the class.[7] Representative generic functions include `print()` and `summary()`. The `print()` function formats the contents of the object assigned to the function simply according to the class of the object, and then displays the formatted contents. For example, the output from `lm()` is divided into several parts, the structure of which can be clearly seen using `str()` (some of the output has been omitted):

```
R> str(lmout)
List of 12
 $ coefficients : Named num [1:3] -1.08 2.11 1.51
  ..- attr(*, "names")= chr [1:3] "(Intercept)" "x1" "x2"

 $ residuals    : Named num [1:10] -0.289 -0.3958 -0.7235 ...
  ..- attr(*, "names")= chr [1:10] "1" "2" "3" "4" ...
 $ effects      : Named num [1:10] -87.911 19.093 -24.186 ...
  ..- attr(*, "names")= chr [1:10] "(Intercept)" "x1" "x2" ...
 $ rank         : int 3
 $ fitted.values: Named num [1:10] 11.3 13.4 32.7 30 25.8 ...
  ..- attr(*, "names")= chr [1:10] "1" "2" "3" "4" ...
 $ assign       : int [1:3] 0 1 2

 ...

 - attr(*, "class")= chr "lm"
```

When applying `print()` to an object of class "`lm`", only some of the core contents are displayed:

---

[7]Although there are two categories of classes, this book focuses only on S3 (S version 3) classes. See the help for package **methods** for S4 (S version 4) classes.

```
R> print(lmout)
Call:
lm(formula = y ~ x1 + x2, data = lmdata)

Coefficients:
(Intercept)           x1           x2
      -1.08         2.11         1.51
```

The same results are obtained by typing the name of the object:

```
R> lmout
Call:
lm(formula = y ~ x1 + x2, data = lmdata)

Coefficients:
(Intercept)           x1           x2
      -1.08         2.11         1.51
```

It should be noted that the information displayed using `print()` is actually part of the contents of the output from functions manipulating statistical models: users who are not familiar with this feature may be under the misapprehension that R's statistical model analysis does not return informative results.

The generic function `summary()` is used to summarize the object assigned to the function. As an example, applying the function to `lmout`, which is an object of class "`lm`", gives:

```
R> summary(lmout)
Call:
lm(formula = y ~ x1 + x2, data = lmdata)

Residuals:
   Min     1Q Median     3Q    Max
-0.899 -0.369 -0.105  0.265  1.192

Coefficients:
            Estimate Std. Error t value Pr(>|t|)
(Intercept)  -1.0770     0.7555   -1.43      0.2
x1            2.1068     0.0851   24.77  4.5e-08
x2            1.5114     0.0477   31.71  8.0e-09

Residual standard error: 0.763 on 7 degrees of freedom
Multiple R-squared:  0.996,     Adjusted R-squared:  0.995
F-statistic:  816 on 2 and 7 DF,  p-value: 5.09e-09
```

5.7.3 Creating dummy variables

Before conducting statistical model analysis, it is sometimes necessary to create dummy variables from factor variables. This operation is easily carried out using the `model.matrix()` function. The `model.matrix()` function is designed to create a design or model matrix (a matrix containing independent variables and a constant). One can also use `model.matrix()`, with some tricks, to generate dummy variables.

A typical call to the `model.matrix()` function for creating dummy variables is given by:

```
R> model.matrix(~x-1, data = obj)
```

where `x` is the factor variable for which the dummy variables are created and `obj` is a data frame object containing `x`. The first argument of `model.matrix()` is specified as a formula type: `-1` is needed to exclude a constant term in the output.

As an example, consider generating dummy variables for one of the variables in `lmdata`. First, a two-level factor variable `x2f` is created from variable `x2` in `lmdata` using the `cut()` function as follows:

```
R> lmdata$x2f <- cut (x = lmdata$x2, breaks = 2,
                      labels = c("a", "b"))
```

where argument `x` is the vector to be converted into a factor; argument `breaks` given the number of interval points to be cut; and argument `labels` denotes the names of the resulting factor levels. Then, the following line of code creates dummy variables for `x2f`, which are concatenated column-wise with `lmdata`:

```
R> lmdata <- cbind(lmdata, model.matrix(~x2f-1, data=lmdata))
```

The resultant `lmdata` includes two dummy variables in the last two columns as follows:

```
R> head(lmdata)
   y x1 x2 x2f x2fa x2fb
1 11  3  4   a    1    0
2 13  4  4   a    1    0
3 32  6 14   b    0    1
4 30  9  8   a    1    0
5 27  2 15   b    0    1
6 33  9 10   a    1    0
```

The above method can be applied to factor variables with three or more levels. The number of created dummy variables is equal to the number of levels. It should be noted that one of the dummy variables must always be omitted in regression analysis to avoid singularity of the design matrix.

The `model.matrix()` function generates dummy variables for factor variables only. "Factor" variables with integer elements could be interpreted by R as vectors of integers. The `model.matrix()` function fails to create dummy variables correctly in this case. For example,

```
R> a <- c(1, 3, 3, 6, 2)
R> model.matrix(~a-1)
  a
1 1
2 3
3 3
4 6
5 2
attr(,"assign")
[1] 1
```

The code above certainly does not create what was intended. To achieve the desired outcome, the "factor" variable must be transformed by the `factor()` function, that is,

```
R> a <- factor(a, levels = 1:6)
R> model.matrix(~a-1)
  a1 a2 a3 a4 a5 a6
1  1  0  0  0  0  0
2  0  0  1  0  0  0
3  0  0  1  0  0  0
4  0  0  0  0  0  1
5  0  1  0  0  0  0
attr(,"assign")
[1] 1 1 1 1 1 1
attr(,"contrasts")
attr(,"contrasts")$a
[1] "contr.treatment"
```

5.7.4 Updating the model

After estimating a model, it is sometimes necessary to modify and then reestimate the model. Such modification and reestimation can easily be conducted using the `update()` function:

```
R> update(object, formula)
```

where argument `object` is the estimated model (old model), output by the model functions, and argument `formula` is the model formula corresponding to the modified model (new model).

Let us consider the situation in which we have to modify the estimated model, `y ~ x1 + x2`, included in `lmout`. For example, instead of variable `x2`, the square of `x2` is to be used as an independent variable in the new model: that is, its formula is `y ~ x1 + I(x2^2)`. Although the new model can be estimated using `lm()` with the new formula and the dataset (`lmdata`), it can also be estimated using `update()` as follows:

```
R> lmout.up <- update(lmout, .~. - x2 + I(x2^2))
```

where `.~.` denotes that the old model formula is used as the model formula in the new model; symbol `-`/`+` denotes that the variable after the symbol is removed from/added to the old model formula; and `I()` is a function that enables us to use the result of the calculation expressed in the parentheses as a variable in the model formula. Therefore, the code given above means that the new model formula consists of the (dependent and independent) variables in the old model included in `lmout`, but with `x2` replaced by the square of `x2`. The reestimated model is given as:

```
R> lmout.up
Call:
lm(formula = y ~ x1 + I(x2^2), data = lmdata)

Coefficients:
(Intercept)           x1      I(x2^2)
     5.3948       2.2612       0.0636
```

5.8 Drawing figures

Although R has powerful functions for drawing graphics, this book uses only the `plot()`, `barplot()`, and `hist()` functions in the **graphics** package. The `plot()` function draws a scatter plot of the two variables assigned to arguments `x` and `y`. The `barplot` function draws a bar plot according to the object assigned to argument `height`. The `hist` function draws a histogram of the values assigned to argument `x`. There are also additional common arguments: for example, `main`, `xlab`, and `ylab` denote the main graph title and the x- and y-axes labels, respectively; and `xlim` and `ylim` are the limits of the x- and y-axes, respectively. Since these functions belong to a generic function, each function provides different types of graphs modified according to the classes of the objects assigned to each function. Each function is briefly explained below.

As an example, four graphs are drawn on one plot region (Figure 5.1). After the plot region has been divided into four sub-regions (2 rows $\times$ 2 columns):

```
R> par(mfrow = c(2, 2))
```

each of the four graphs is drawn. The first figure, which is located in the top-left of the region, is a scatter plot of `x1` and `y` in `lmdata`:

```
R> plot(x = lmdata$x1,
        y = lmdata$y,
        main = "Scatter plot of x1 and y in lmdata",
        xlim = c(0, 10),
        ylim = c(0, 50))
```

The second figure, which is located in the top-right of the region, is a histogram of `x1` in object `lmdata`:

```
R> hist(x = lmdata$x1)
```

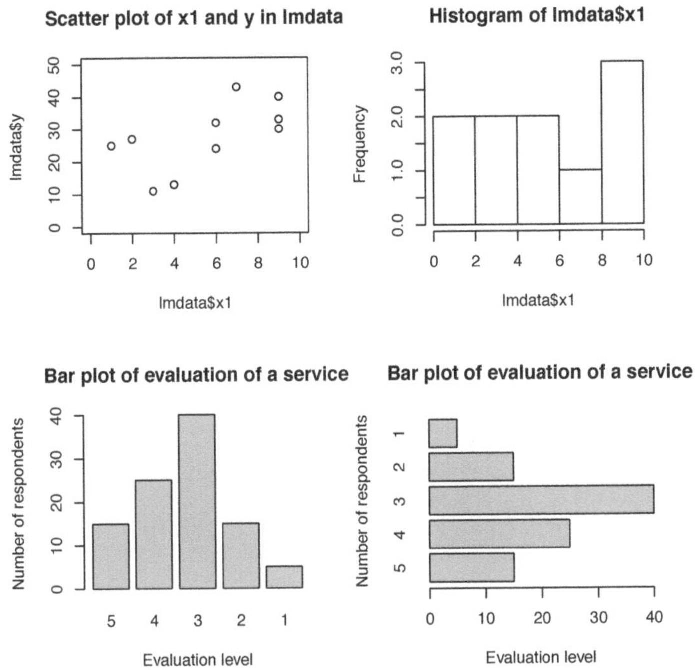

Figure 5.1 *Examples of* `plot()`, `barplot()`, *and* `hist()`.

The third (bottom-left) and fourth (bottom-right) graphs are bar plots. Suppose that 100 respondents evaluate a service according to a five-point scale (with 5 being the most preferred and 1 being the least preferred). The data is stored in a vector `evaluation` together with the names of the five points. We then create bar plots of `evaluation` with the bars drawn both vertically and horizontally as follows:

```
R> evaluation <- c(15, 25, 40, 15, 5)
R> names(evaluation) <- c(5, 4, 3, 2, 1)
R> barplot(height = evaluation,
           main = "Bar plot of evaluation of a service",
           xlab = "Evaluation level",
           ylab = "Number of respondents")
R> barplot(height = evaluation,
           main = "Bar plot of evaluation of a service",
           xlab = "Evaluation level",
           ylab = "Number of respondents",
           horiz = TRUE) # draw the bars horizontally
```

Figures created by `plot()`, `barplot()`, `hist()`, among others can be exported as (Windows) Metafile, (Encapsulated) PostScript, PDF, PNG, bitmap, TIFF, or JPG format for use in other documents or presentations. Methods for saving figures can be practically classified into two. In the first method, the figure is drawn on the screen and then saved in the desired format, while the other writes the figure directly into a file in the desired format. The former method can be carried out through the menu bar in the R console window, whereas the latter is realized by appropriate R functions such as `pdf()`, `postscript()`, `win.metafile()` (for Windows only). Please refer to the corresponding help file for the details.

Appendix A

Other Packages Related to This Book

A.1 Introduction

This appendix gives additional information on R packages related to the topics covered in the previous chapters, namely, contingent valuation (CV), discrete choice models, cluster analysis, principal component analysis, factor analysis, and conjoint analysis. However, our intent is not to provide comprehensive information about each of the related packages; readers who are interested in the following analyses in R, should consult websites and/or books related to R.

A.2 Contingent valuation

While the single-bounded/double-bounded dichotomous choice format has mainly been applied to empirical CV studies, a payment card format has occasionally also been used (e.g., Nordström, 2012; Hu et al., 2011). For a question in the payment card format, a candidate set of willingness-to-pay (WTP) values, expressed in interval values, is created and respondents are requested to select a card (alternative) containing their actual WTP. Packages **iwtp** (Belyaev and Zhou, 2012) and **intReg** (Toomet, 2012) are available for nonparametric and parametric approaches for conducting statistical analysis of responses to payment card CV questions.

The `iwtp()` function is the main one in the **iwtp** package and can be used to estimate mean WTP values. The basic usage is given as:

```
iwtp(data, dist)
```

where argument `data` is a dataset containing the upper and lower bounds selected by the respondents and argument `dist` is a type of survival function (Weibull or mixed Weibull and exponential distributions). This function returns parametric estimation of WTPs based on the survival function set by the user and nonparametric estimation of WTPs based on the empirical survival function. Furthermore, the `resam()` function in **iwtp** is provided to calculate the confidence intervals of parametric WTP estimates using a re-sampling approach.

The `intReg()` function in the **intReg** package can be used to estimate interval regression models and supports four types ofdistributions (normal, logistic, log-log, and Cauchy). The basic usage is given as:

```
intReg(formula, data)
```

where argument `formula` is the model formula and argument `data` is a dataset. The format of `formula` is similar to that in `lm()`, where the left-hand side of the tilde contains the dependent variable, and the right-hand side comprises the independent variables (e.g., `y ~ x1 + x2`).

A.3 Discrete choice models

Various R packages provide functions for carrying out discrete choice model analysis. As an example, the **mlogit** package (Croissant, 2012) is a powerful and useful package that specializes in discrete choice models. It can estimate basic and advanced discrete choice models using the `mlogit()` function. The basic usage of the function for estimating a conditional or multinomial logit model is given simply as:

```
mlogit(formula, data)
```

Although it is necessary to set `formula` according to the types of the independent variables (i.e., an alternative- or individual-specific variable), the basic format of `formula` is similar to that in `lm()`. To estimate advanced discrete choice models, additional arguments are merely appended to the above. For example, when a multinomial probit model is estimated, argument `probit` is set to `TRUE`; when a heteroscedastic logit model is estimated, argument `heterosc` is set to `TRUE`; and when a random parameter logit model is estimated, argument `rpar` is set according to the user's assumption. Readers who are interested in this package should refer to the reference manual and informative vignettes.

As another example, the **bayesm** package (Rossi, 2012) estimates discrete choice models based on Bayesian methods. It provides functions for estimating binary/multinomial probit, multinomial logit, and hierarchical binary/multinomial logit models. Readers who are interested in this package should refer to the reference manual and Rossi et al. (2005).[1]

These packages may be useful for conducting advanced discrete choice experiments, but this is beyond the scope of this book. Readers should note that it may not be possible to construct an appropriate design for the above-mentioned models using the design methods explained in this book. Therefore, those wishing to conduct an advanced discrete choice experiment are advised to search for a design method that is appropriate for the chosen discrete choice model they wish to apply and to construct the design based on the given design method.

---

[1]Other packages **nnet** (Venables and Ripley, 2002), **VGAM** (Yee and Wild, 1996; Yee, 2012), **Zelig** (Imai et al., 2008, 2012), and **MNP** (Imai and van Dyk, 2011) also provide functions related to discrete choice models. Kleiber and Zeileis (2008), Aitkin et al. (2009), and Tutz (2011) explain discrete choice models and the relevant microeconometric methods in R.

A.4 Cluster, component, and factor analysis

Before analyzing stated preference (SP) data, other statistical methods are frequently used. For example, respondents may be classified into two or more groups on the basis of responses to questions other than the SP questions, and/or independent variables used in the SP analysis may be created from responses to questions that are expected to affect the responses to the SP questions. Cluster analysis, principal component analysis, and exploratory factor analysis are used in such preparations.

The `hclust()` function in the **stats** package (R Core Team, 2014) can be used to implement hierarchical cluster analysis. The basic usage is given as:

```
hclust(d, method)
```

where argument `d` is a distance matrix created using the `dist()` function based on a distance measure set by the user (e.g., Euclidean distance or maximum distance) and a dataset, whereas argument `method` is a method for clustering individuals (e.g., the Ward or single linkage method).

If hierarchical cluster analysis is not suitable (i.e., for clustering based on a larger dataset), partitioning cluster analysis is available. For example, k-means clustering and a robust version of this method can be implemented using the `kmeans()` function in the **stats** package and the `pam()` function in the **cluster** package (Maechler et al., 2012), respectively. The basic usage of these functions is given as:

```
kmeans(x, centers)
pam(x, centers)
```

where argument `x` is a dataset and argument `centers` is the number of clusters. See Section 4.4.2 for an example of applying `kmeans()`.

Principal component analysis and exploratory factor analysis[2] can be implemented using functions `principal()` and `fa()` in the **psych** package (Revelle, 2012),[3] respectively. The basic usage of these functions is given as:

```
principal(r, nfactors, rotate, scores)
fa(r, nfactors, rotate, scores, fm)
```

where argument `r` is a correlation or raw data matrix, argument `nfactors` is the number of components/factors to be extracted, argument `rotate` is a method for rotating/transforming the solution, argument `scores` is set to `TRUE` if component/factor scores need to be calculated based on the result, and

[2] A good alternative approach for conducting these analyses is structural equation modeling, which can be implemented in R using the **sem** package (Fox et al., 2012).

[3] The **psych** package provides valuable functions for psychological research and seems to be useful for analyzing data gathered through questionnaire surveys conducted in other disciplines. For example, it provides the `read.clipboard()` function for reading a dataset contained in a clipboard into R, the `alpha()` function, which calculates Cronbach's α used to measure internal consistency of reliability for a single scale, and the `corr.test()` function, which calculates a correlation matrix and the p-value corresponding to each value therein. Refer to the reference manual and vignettes for the details.

argument `fm` is the factoring method to be used. When creating independent variables for SP analysis, `scores` must be set to `TRUE`.[4]

A.5 Conjoint analysis

Although conjoint analysis is beyond the scope of this book, it has been widely applied in various fields by practitioners and academicians (e.g., Green et al., 2004; Orme, 2010). Thus, for readers who are interested in conjoint analysis, the **conjoint** package (Bak and Bartlomowicz, 2012) is introduced. It contains a variety of functions—from creating profiles to extracting valuable information from the results of the analysis. Some of these are presented below.

The `caFactorialDesign()` function creates a set of profiles based on a full or fractional factorial design by internally using the `optFederov()` function in the **AlgDesign** package (Wheeler, 2014). The basic usage is given as:

```
caFactorialDesign(data, type, cards)
```

where argument `data` contains the attributes and their levels, argument `type` is the design method (e.g., full factorial or fractional factorial), and argument `cards` is the number of profiles.

The resultant design can be converted into an integer value format suited to the post-processing functions (i.e., conducting statistical analysis or calculating the relative importance of each level) using the `caEncodedDesign()` function:

```
caEncodedDesign(design)
```

where argument `design` is the resultant design.

The `caModel()` function conducts a regression analysis of conjoint data, consisting of the encoded design and responses to profiles (the respondent dataset), by internally using the `lm()` function in the **stats** package and returns the estimates. The basic usage is given as:

```
caModel(y, x)
```

where argument `y` is the respondent dataset and argument `x` is the encoded design.

Other useful functions such as those for calculating partial utilities of levels for each respondent and for predicting market share of profiles, are also included in the package; see the reference manual of the package for details.

---

[4]The **stats** package also includes the `princomp()` and `prcomp()` functions for principal component analysis and the `factanal()` function for factor analysis.

Appendix B

Examples of Contrivance in Empirical Studies

B.1 Introduction

Empirical studies generally aim to extract valuable insights that can be used to grapple with target issues on the basis of empirical analysis results. However, there is no single method that is suitable for all issues; each method has its own strengths and weaknesses. Therefore, it is important to select a method that is suited to a particular target issue to make the best use of the strengths of the method and to offset the weaknesses of the method by jointly using other methods or techniques. This also covers empirical stated preference (SP) studies. A way of improving a user's ability to apply SP methods to his/her own empirical studies is to understand the various contrivances in the application of SP methods in previous empirical studies. In addition to empirical studies listed in the introductory sections of Chapters 2, 3, and 4, this appendix briefly presents some examples of contrivances in previous empirical SP studies.

B.2 Providing information to respondents

A strength of SP methods is the ability to control the situations in which respondents answer SP questions; various studies have examined the effects of providing information on the respondents' answers to the SP questions. There are two ways to test such information effects.

The first is a within-subjects test, in which the respondents are asked identical questions before and after the information has been received; the information effect is tested by comparing the differences in the answers. For example, Alfnes et al. (2006) conducted a discrete choice experiment (DCE) study on consumers' valuations of salmon fillets and compared the respondents' preferences for the color of the fish before and after receiving information about the origin of the color. Rousseau and Vranken (2013) asked the respondents in their study to evaluate apple alternatives six times in a DCE study before providing information about the cultivation method of apples.

The respondents were then asked whether they would change their responses in the six choice tasks after being given the information.

The other way of testing the information effect is a between-subjects test, in which the respondents are divided into two or more groups, and each group is given different information and then requested to answer identical questions. The information effect is tested by comparing the different groups' responses to the questions. For example, Corso et al. (2001) applied contingent valuation (CV) to estimate willingness-to-pay (WTP) to reduce automobile-related mortality risk by using a side-impact airbag. The respondents were divided into four groups, and the risk was explained to each group in a different way. The researchers then examined the differences in WTP values among the four groups. Alberini et al. (2005) used CV to examine whether respondents' answers to CV questions regarding a conservation plan for a cultural heritage site were affected by providing them with information about the reasons why others agreed or disagreed with the plan through a between-subjects test.

We can consider changing the context of choice situations as a way of providing information. For example, using DCEs Jaeger and Rose (2008) examined the effects of consumption situations on the behavior of consumers in choosing fruit. They constructed situations in which the respondents were asked to select their most preferred fruit from a set of fruits based on six situational, social, and emotional components such as location and mental state. Molin and Timmermans (2010) conducted a DCE study to test the effect of choice situations, expressed as seven components, including purpose of the trip, weather, and baggage, on the respondents' choice of travel mode from a station to their final destination. Casini et al. (2009) compared consumers' preferences for wine in two different situations—a meal with friends and a dinner at home with friends—estimated using best–worst scaling (BWS).

Although the most popular medium for providing information seems to be text, pictures, figures, and movies have also been used. For example, Corso et al. (2001) used visual aids to explain mortality risks, while Kling et al. (2004), who conducted a CV study to evaluate a local historic hotel, used a photograph to explain the features of the target hotel. Araña and León (2009) divided their respondents into three groups. Each group watched a different video clip that made them feel disgusted, sad, or neutral, and then each group was asked to answer DCE questions regarding a plan to improve the environmental burdens of a stone mining facility.

B.3 Using product/service samples

A more positive intervention than the provision of information is to request respondents for a trial of product/service samples related to the objective of the study. For example, Laver et al. (2011) examined the possibility of using Nintendo's Wii Fit for physical therapy by means of a within-subjects test. They conducted the same DCE survey measuring the patients' preferences for the Wii Fit before and after applying physical therapy with the Wii Fit and then examined the differences in WTP values before and after the physical therapy.

Another way of using trials is to request respondents to taste food/drink samples. Although tasting may increase the burden placed on the respondents to answer the SP questions, it can diminish ambiguity in the responses because respondents would be able to recognize their preferences more clearly. For example, Enneking et al. (2007) conducted a DCE study in which a soft drink was treated as an alternative, with attributes including sweetness, brand name, calorie content, and price. In order to categorize the drink's sweetness, the respondents were asked to taste it. Jaeger et al. (2008) applied BWS to elicit the preferences of consumers for minced pork patties by asking the respondents to taste the pork samples. Thomson et al. (2010) used BWS to investigate appropriate words to express consumers' emotions when eating chocolate. The respondents were asked to select the most suitable and least suitable words to describe the chocolates after tasting each one.[1]

B.4 Cost–benefit analysis and valuation

Since the primary objective of developing CV is to estimate the benefits used in cost–benefit analysis related to the environment, an idea that combines the SP methods with cost–benefit analysis (or other project/policy valuation techniques) has been observed in various fields (Carson, 2012). For example, Holmes et al. (2004) conducted cost–benefit analysis of the riparian ecosystem restoration projects for a watershed based on the ecosystem benefits estimated using the CV and found that the benefits were three or more times greater than the costs of riparian restoration projects. Haefeli et al. (2008) estimated the benefits of three types of spinal surgery using CV, compared the benefits with the actual costs of three types of surgery, and found that the benefits were greater than the costs in two of the three types, but less than the costs in the third type. Buzby et al. (1995) carried out a cost–benefit study related to a food safety policy: the banning of a post-harvest pesticide used to reduce the loss of grapefruit. They applied CV to estimate the benefits to consumers of a ban that would reduce the risk of cancer related to exposure to the pesticide and found that the benefits of the ban were greater than the costs. Fleischer and Felsenstein (2002) conducted a cost–benefit analysis of a cultural event—televising a famous music contest—that includes the public benefit from viewing the televised event estimated using CV.

However, there has been some criticism that the estimation of benefits using the SP methods could be restrictive for respondents who are not familiar with the target issue. They may feel that they have to answer the SP questions without sufficient opportunity to examine the issue, gather additional

[1] Actual products are also used in a non-hypothetical SP approach, where the respondents are requested to actually purchase a product corresponding to an alternative they would like to purchase based on hypothetical (general) SP questions (e.g., Lusk and Schroeder, 2004).

information about the issue, and discuss the issue with others. To overcome this limitation, a deliberative monetary valuation approach—also known as a workshop-based, citizens' jury, market stall, or participatory approach—has been introduced into SP studies, where small groups of participants evaluate the issue through deliberation (Macmillan et al., 2002). For example, Macmillan et al. (2002) compared the benefits of the conservation of a wild animal estimated using both a conventional and deliberative CV approach. In the latter approach, the participants were requested to attend two hour-long meetings held one week apart. Roosen et al. (2011) used a CV approach to examine the effect of providing information and an opportunity to discuss matters related to nanotechnology on the consumers' valuation of a juice supplemented with vitamin D through nanotechnology.

B.5 Using SP study results in simulations

Simulation using the results of the SP methods is a classic application, especially in transportation research. Transportation research has developed simulation systems for forecasting the various travel behaviors of commuters (e.g., Ben-Akiva and Lerman, 1985; Train, 1986). In such systems, the results of SP studies have been used widely. For example, Takama and Preston (2008) developed an agent-based simulation system to examine the effect of a road use charge on peoples' behavior in visiting an area. The system included a mode choice model estimated using SP data as well as a parking location choice model estimated using revealed preference data. Further, using the results of SP studies, Hunt and Stefan (2007) developed a system to simulate urban commercial movement.

Various other simulation techniques have also used the results of SP studies. For example, Mallawaarachchi and Quiggin (2001) constructed a framework to evaluate economically and environmentally, the regional land resource management alternatives on the basis of mathematical programming. They integrated the environmental benefits of preserving the land from agricultural usage—estimated using a DCE study—into the framework and then applied the integrated approach to evaluate land management for sugar cane cultivation. Nam et al. (2010) developed a computable general equilibrium model for measuring and predicting the effects of air pollution on the European economy. Various economic and demographic data were used in their system, in which a part of the costs relating to air pollution were estimated using SP methods. Bateman et al. (2003) created a geographical information system for cost–benefit analysis of land resource management from an environmental economics perspective. The system included benefits related to land use derived from applying the CV approach as well as the travel cost method. Rivers and Jaccard (2005) used the estimates from a DCE study that analyzed plant managers' behaviors in selecting energy generation systems in an attempt to model the interactions of energy technologies with the economy.

Bibliography

Aabø S (2005). "Valuing the benefits of public libraries." *Information Economics and Policy*, **17**(2), 175–198.

Adamowicz W, Boxall P, Williams M, Louviere J (1998). "Stated preference approaches for measuring passive use values: Choice experiments and contingent valuation." *American Journal of Agricultural Economics*, **80**(1), 64–75.

Adamowicz W, Louviere J, Williams M (1994). "Combining revealed and stated preference methods for valuing environmental amenities." *Journal of Environmental Economics and Management*, **26**, 271 – 292.

Aitkin M, Francis B, Hinde J, Darnell R (2009). *Statistical Modelling in R*. Oxford University Press, Oxford, UK.

Aizaki H (2012a). "Basic functions for supporting an implementation of choice experiments in R." *Journal of Statistical Software, Code Snippets*, **50**(2), 1–24. URL `http://www.jstatsoft.org/v50/c02/`.

Aizaki H (2012b). ***mded**: Measuring the Difference between Two Empirical Distributions, R Package Version 0.1-0.* URL `http://CRAN.R-project.org/package=mded`.

Aizaki H (2014a). ***support.BWS**: Basic Functions for Supporting an Implementation of Best–Worst Scaling, R Package Version 0.1-1.* URL `http://CRAN.R-project.org/package=support.BWS`.

Aizaki H (2014b). ***support.CEs**: Basic Functions for Supporting an Implementation of Choice Experiments, R Package Version 0.3-1.* URL `http://CRAN.R-project.org/package=support.CEs`.

Alberini A (1995). "Optimal designs for discrete choice contingent valuation surveys: Single-bound, double-bound, and bivariate models." *Journal of Environmental Economics and Management*, **28**(3), 287–306.

Alberini A, Rosato P, Longo A, Zanatta V (2005). "Information and willingness to pay in a contingent valuation study: The value of S. Erasmo in the Lagoon of Venice." *Journal of Environmental Planning and Management*, **48**(2), 155–175.

Alfnes F (2004). "Stated preferences for imported and hormone-treated beef: Application of a mixed logit model." *European Review of Agricultural Economics*, **31**(1), 19–37.

Alfnes F, Guttormsen AG, Steine G, Kolstad K (2006). "Consumers' willingness to pay for the color of salmon: A choice experiment with real economic incentives." *American Journal of Agricultural Economics*, **88**(4), 1050–1061.

Amaya-Amaya M, Gerard K, Ryan M (2008). "Discrete choice experiments in a nutshell." In M Ryan, K Gerard, M Amaya-Amaya (eds.), *Using Discrete Choice Experiments to Value Health and Health Care*, pp. 13–46. Springer, Dordrecht, NL.

Araña JE, León CJ (2009). "Understanding the use of non-compensatory decision rules in discrete choice experiments: The role of emotions." *Ecological Economics*, **68**(8–9), 2316–2326.

Asrat S, Yesuf M, Carlsson F, Wale E (2010). "Farmers' preferences for crop variety traits: Lessons for on-farm conservation and technology adoption." *Ecological Economics*, **69**(12), 2394–2401.

Atkinson G, Healey A, Mourato S (2005). "Valuing the costs of violent crime: A stated preference approach." *Oxford Economic Papers*, **57**(4), 559–585.

Auger P, Devinney TM, Louviere JJ (2007). "Using best–worst scaling methodology to investigate consumer ethical beliefs across countries." *Journal of Business Ethics*, **70**, 299–326.

Bak A, Bartlomowicz T (2012). ***conjoint**: Conjoint Analysis Package, R Package Version 1.35.* URL `http://CRAN.R-project.org/package=conjoint`.

Bateman IJ, Carson RT, Day B, Hanemann M, Hanley N, Hett T, Jones-Lee M, Loomes G, Mourato S, Özdemiroglu E, Pearce DW, Sugden R, Swanson J (eds.) (2002). *Economic Valuation with Stated Preference Techniques: A Manual.* Edward Elgar, Cheltenham, UK.

Bateman IJ, Lovett AA, Brainard JS (2003). *Applied Environmental Economics: A GIS Approach to Cost–Benefit Analysis.* Cambridge University Press, Cambridge, UK.

Bedate AM, Herrero LC, Sanz JÁ (2009). "Economic valuation of a contemporary art museum: Correction of hypothetical bias using a certainty question." *Journal of Cultural Economics*, **33**(3), 185–199.

Belyaev Y, Zhou W (2012). ***iwtp**: Numerical Evaluation Willingness to Pay Based on Interval Data, R Package Version 1.0-0.* URL `http://CRAN.R-project.org/package=iwtp`.

Ben-Akiva M, Lerman SR (1985). *Discrete Choice Analysis: Theory and Application to Travel Demand.* The MIT Press, MA, USA.

Bennett J, Adamowicz V (2001). "Some fundamentals of environmental choice modelling." In J Bennett, R Blamey (eds.), *The Choice Modelling Approach to Environmental Valuation*, pp. 37–69. Edward Elgar, Cheltenham, UK.

Bennett J, Birol E (eds.) (2010). *Choice Experiments in Developing Countries: Implementation, Challenges and Policy Implications.* Edward Elgar, Cheltenham, UK.

Bennett J, Blamey R (eds.) (2001). *The Choice Modelling Approach to Environmental Valuation.* Edward Elgar, Cheltenham, UK.

Bennett J, Rolfe J, Morrison M (2001). "Remnant vegetation and wetlands protection: Non-market valuation." In J Bennett, R Blamey (eds.), *The Choice Modelling Approach to Environmental Valuation*, pp. 93–114. Edward Elgar, Cheltenham, UK.

Bergstrom JC, Ready RC (2009). "What have we learned from over 20 years of farmland amenity valuation research in North America?" *Review of Agricultural Economics*, **31**, 21–49.

Bernabéu R, Díaz M, Olivas R, Olmeda M (2012). "Consumer preferences for wine applying best–worst scaling: A Spanish case study." *British Food Journal*, **114**(9), 1228–1250.

Bhattacharya S, Alberini A, Cropper ML (2007). "The value of mortality risk reductions in Delhi, India." *Journal of Risk and Uncertainty*, **34**(1), 21–47.

Birol E, Koundouri P (eds.) (2008). *Choice Experiments Informing Environmental Policy: A European Perspective.* Edward Elgar, Cheltenham, UK.

Bishai DM, Lang HC (2000). "The willingness to pay for wait reduction: The disutility of queues for cataract surgery in Canada, Denmark, and Spain." *Journal of Health Economics*, **19**(2), 219–230.

Bishop RC, Heberlein TA (1979). "Measuring values of extramarket goods: Are indirect measures biased?" *American Journal of Agricultural Economics*, **61**(5), 926–930.

Bliemer MCJ, Rose JM (2011). "Experimental design influences on stated choice outputs: An empirical study in air travel choice." *Transportation Research Part A: Policy and Practice*, **45**(1), 63–79.

Boccaletti S, Nardella M (2000). "Consumer willingness to pay for pesticide-free fresh fruit and vegetables in Italy." *The International Food and Agribusiness Management Review*, **3**(3), 297–310.

Bockstael NE, Freeman III AM (2005). "Welfare theory and valuation." In KG Mäler, JR Vincent (eds.), *Handbook of Environmental Economics*, volume 2, chapter 12, pp. 517–570. Elsevier, New York.

Boman M, Bostedt G (1999). "Valuing the wolf in Sweden: Are benefits contingent on the supply?" In M Boman, R Brännlund, B Kriström (eds.), *Topics in Environmental Economics*, chapter 9, pp. 157–174. Kluwer Academic Publishers, Boston.

Boman M, Bostedt G, Kriström B (1999). "Obtaining welfare bounds in discrete-response valuation studies: A non-parametric approach." *Land Economics*, **75**(2), 284–294.

Boxall PC, Adamowicz WL, Swait J, Williams M, Louviere J (1996). "A comparison of stated preference methods for environmental valuation." *Ecological Economics*, **18**, 243 – 253.

Boyle KJ (2003). "Contingent valuation in practice." In PA Champ, KJ Boyle, TC Brown (eds.), *A Primer on Nonmarket Valuation*, chapter 5, pp. 111–169. Kluwer Academic Publishers, Boston.

Boyle KJ, Welsh MP, Bishop RC (1988). "Validation of empirical measures of welfare change: Comment." *Land Economics*, **64**(1), 94–98.

Brooks K, Lusk JL (2010). "Stated and revealed preferences for organic and cloned milk: Combining choice experiment and scanner data." *American Journal of Agricultural Economics*, **92**(4), 1229–1241.

Burr JM, Kilonzo M, Vale L, Ryan M (2007). "Developing a preference-based glaucoma utility index using a discrete choice experiment." *Optometry & Vision Science*, **84**(8), E797–E809.

Buzby JC, Ready RC, Skees JR (1995). "Contingent valuation in food policy analysis: A case study of a pesticide-residue risk reduction." *Journal of Agricultural and Applied Economics*, **27**(2), 613–625.

Cameron AC, Trivedi PK (2005). *Microeconometrics: Methods and Applications*. Cambridge University Press, New York.

Cameron TA (1988). "A new paradigm for valuing non-market goods using referendum data: Maximum likelihood estimation by censored logistic regression." *Journal of Environmental Economics and Management*, **15**(3), 355–379.

Cameron TA, James MD (1987). "Efficient estimation methods for 'closed-ended' contingent valuation surveys." *The Review of Economics and Statistics*, **69**(2), 269–276.

Carlsson F, Frykblom P, Lagerkvist CJ (2007). "Consumer benefits of labels and bans on GM foods—Choice experiments with Swedish consumers." *American Journal of Agricultural Economics*, **89**(1), 152–161.

Carlsson F, Johansson-Stenman O, Martinsson P (2004). "Is transport safety more valuable in the air?" *Journal of Risk and Uncertainty*, **28**(2), 147–163.

Carson RT (1985). *Three Essays on Contingent Valuation*. Dissertation, University of California Berkeley.

Carson RT (2012). *Contingent Valuation: A Comprehensive Bibliography and History*. Edward Elgar, Northampton, MA.

Carson RT, Hanemann WM (2005). "Contingent valuation." In KG Mäler, JR Vincent (eds.), *Handbook of Environmental Economics*, volume 2, chapter 17, pp. 821–936. Elsevier, New York.

Carson RT, Mitchell RC, Hanemann WM, Kopp RJ, Presser S, Ruud PA (1992). *A Contingent Valuation Study of Lost Passive Use Values Re-*

sulting from the Exxon Valdez Oil Spill. Report to the Attorney General of the State of Alaska. Natural Resource Damage Assessment Inc. URL http://mpra.ub.uni-muenchen.de/6984/.

Carson RT, Mitchell RC, Hanemann WM, Kopp RJ, Presser S, Ruud PA (2003). "Contingent valuation and lost passive use: Damages from the *Exxon Valdez oil spill.*" *Environmental and Resource Economics*, **25**(3), 257–286.

Carson RT, Steinberg D (1990). "Experimental design for discrete choice voter preference surveys." In *1989 Proceeding of the Survey Methodology Section of the American Statistical Association*, pp. 821–822.

Casini L, Corsi AM, Goodman S (2009). "Consumer preferences of wine in Italy applying best–worst scaling." *International Journal of Wine Business Research*, **21**(1), 64–78.

Chrysochou P, Corsi AM, Krystallis A (2012a). "What drives Greek consumer preferences for cask wine?" *British Food Journal*, **114**(8), 1072–1084.

Chrysochou P, Krystallis A, Mocanu A, Lewis RL (2012b). "Generation Y preferences for wine: An exploratory study of the US market applying the best–worst scaling." *British Food Journal*, **114**(4), 516–528.

Chrzan K, Golovashkina N (2006). "An empirical test of six stated importance measures." *International Journal of Market Research*, **48**(6), 717–740.

Chrzan K, Orme B (2000). "An overview and comparison of design strategies for choice-based conjoint analysis." *Sawtooth Software Research Paper Series*, pp. 1–18. URL http://www.sawtoothsoftware.com/support/technical-papers.

Chrzan K, Patterson M (2006). "Testing for the optimal number of attributes in MaxDiff questions." *Sawtooth Software Research Paper Series*, pp. 1–7.

Coast J, Flynn TN, Natarajan L, Sproston K, Lewis J, Louviere JJ, Peters TJ (2008). "Valuing the ICECAP capability index for older people." *Social Science & Medicine*, **67**, 874–882.

Coast J, Horrocks S (2007). "Developing attributes and levels for discrete choice experiments using qualitative methods." *Journal of Health Services Research & Policy*, **12**(1), 25–30.

Coast J, Salisbury C, de Berker D, Noble A, Horrocks S, Peters TJ, Flynn TN (2006). "Preferences for aspects of a dermatology consultation." *British Journal of Dermatology*, **155**, 387–392.

Cochran WG, Cox GM (1957). *Experimental Designs.* Second edition. John Wiley & Sons, Inc., Hoboken, New Jersey.

Cohen E (2009). "Applying best–worst scaling to wine marketing." *International Journal of Wine Business Research*, **21**(1), 8–23.

Cohen E, d'Hauteville F, Sirieix L (2009). "A cross-cultural comparison of choice criteria for wine in restaurants." *International Journal of Wine Business Research*, **21**(1), 50–63.

Cohen S, Neira L (2004). "Measuring preference for product benefits across countries: Overcoming scale usage bias with maximum difference scaling." *Excellence in International Research*, pp. 1–22.

Cohen SH (2003). "Maximum difference scaling: Improved measures of importance and preference for segmentation." *Sawtooth Software Research Paper Series*, pp. 1–17. URL `http://www.sawtoothsoftware.com/support/technical-papers`.

Coltman TR, Devinney TM, Keating BW (2011). "Best–worst scaling approach to predict consumer choice for 3PL services." *Journal of Business Logistics*, **32**, 139–152.

Cooper J, Loomis J (1992). "Sensitivity of willingness-to-pay estimates to bid design in dichotomous choice contingent valuation models." *Land Economics*, **68**(2), 211–224.

Corso PS, Hammitt JK, Graham JD (2001). "Valuing mortality-risk reduction: Using visual aids to improve the validity of contingent valuation." *Journal of Risk and Uncertainty*, **23**(2), 165–184.

Cranfield JAL, Magnusson E (2003). "Canadian consumer's willingness-to-pay for pesticide free food products: An ordered probit analysis." *International Food and Agribusiness Management Review*, **6**, 13–30.

Croissant Y (2014). ***Ecdat**: Datasets for Econometrics.* R package version 0.2-4, URL `http://CRAN.R-project.org/package=Ecdat`.

Croissant Y (2012). ***mlogit**: Multinomial Logit Model, R Package Version 0.2-2.* URL `http://CRAN.R-project.org/package=mlogit`.

Cross P, Rigby D, Edwards-Jones G (2012). "Eliciting expert opinion on the effectiveness and practicality of interventions in the farm and rural environment to reduce human exposure to *Escherichia coli* O157." *Epidemiology & Infection*, **140**, 643–654.

Crouch GI, Louviere JJ (2003). "Experimental analysis of the choice of convention site." *Tourism Analysis*, **8**, 171–176.

Curtis KR, Moeltner K (2006). "Genetically modified food market participation and consumer risk perceptions: A cross-country comparison." *Canadian Journal of Agricultural Economics*, **54**(2), 289–310.

Daly TM, Lee JA, Soutar GN, Rasmi S (2010). "Conflict-handling style measurement: A best–worst scaling application." *International Journal of Conflict Management*, **21**(3), 281–308.

de Bekker-Grob EW, Essink-Bot ML, Meerding WJ, Pols HAP, Koes BW, Steyerberg EW (2008). "Patients' preferences for osteoporosis drug treatment: A discrete choice experiment." *Osteoporosis International*, **19**, 1029–1037.

de Bekker-Grob EW, Hol L, Donkers B, Van Dam L, Habbema JDF, Van Leerdam ME, Kuipers EJ, Essink-Bot ML, Steyerberg EW (2010). "Labeled versus unlabeled discrete choice experiments in health economics: An application to colorectal cancer screening." *Value in Health*, **13**(2), 315–323.

de Bekker-Grob EW, Ryan M, Gerard K (2012). "Discrete choice experiments in health economics: A review of the literature." *Health Economics*, **21**(2), 145–172.

Dekhili S, Sirieix L, Cohen E (2011). "How consumers choose olive oil: The importance of origin cues." *Food Quality and Preference*, **22**(8), 757–762.

Delaney L, O'Toole F (2006). "Willingness to pay: Individual or household?" *Journal of Cultural Economics*, **30**, 305–309.

del Saz Salazar S, Marques JM (2005). "Valuing cultural heritage: The social benefits of restoring and old Arab tower." *Journal of Cultural Heritage*, **6**(1), 69–77.

de Magistris T, Groot E, Gracia A, Albisu LM (2011). "Do millennial generation's wine preferences of the 'New World' differ from the 'Old World'?: A pilot study." *International Journal of Wine Business Research*, **23**(2), 145–160.

de Meijer C, Brouwer W, Koopmanschap M, van den Berg B, van Exel J (2010). "The value of informal care: A further investigation of the feasibility of contingent valuation in informal caregivers." *Health Economics*, **19**, 755–771.

Dickie M, Messman VL (2004). "Parental altruism and the value of avoiding acute illness: Are kids worth more than parents?" *Journal of Environmental Economics and Management*, **48**(3), 1146–1174.

Dillman DA (2000). *Mail and Internet Surveys: The Tailored Design Method*. 2nd edition. John Wiley & Sons, NY, USA.

Ding Y, Veeman MM, Adamowicz WL (2012). "The influence of attribute cutoffs on consumers' choices of a functional food." *European Review of Agricultural Economics*, **39**(5), 745–769.

Dong H, Kouyatea B, Cairns J, Mugisha F, Sauerborn R (2003). "Willingness-to-pay for community-based insurance in Burkina Faso." *Health Economics*, **12**, 849–862.

Dranitsaris G, Longo CJ, Grossman LD (2000). "The economic value of a new insulin preparation, Humalog® Mix 25™: Measured by a willingness-to-pay approach." *PharmacoEconomics*, **18**(3), 275–287.

Duffield JW, Patterson DA (1991). "Inference and optimal design for a welfare measure in dichotomous choice contingent valuation." *Land Economics*, **67**(2), 225–239.

Enneking U (2004). "Willingness-to-pay for safety improvements in the German meat sector: The case of the Q&S label." *European Review of Agricultural Economics*, **31**(2), 205–223.

Enneking U, Neumann C, Henneberg S (2007). "How important intrinsic and extrinsic product attributes affect purchase decision." *Food Quality and Preference*, **18**(1), 133–138.

Erdem S, Rigby D, Wossink A (2012). "Using best–worst scaling to explore perceptions of relative responsibility for ensuring food safety." *Food Policy*, **37**, 661–670.

Evans G (1999). "The economics of the national performing arts- Exploiting consumer surplus and willingness-to-pay: A case of cultural policy failure?" *Leisure Studies*, **18**(2), 97–118.

Farrar S, Ryan M, Ross D, Ludbrook A (2000). "Using discrete choice modelling in priority setting: An application to clinical service developments." *Social Science & Medicine*, **50**(1), 63–75.

Fay MP, Shaw PA (2010). "Exact and asymptotic weighted logrank tests for interval censored data: The **interval** R Package." *Journal of Statistical Software*, **36**(2), 1–34. URL `http://www.jstatsoft.org/v36/i02/`.

Ferrini S, Scarpa R (2007). "Designs with *a priori* information for nonmarket valuation with choice experiments: A Monte Carlo study." *Journal of Environmental Economics and Management*, **53**, 342–363.

Finn A, Louviere JJ (1992). "Determining the appropriate response to evidence of public concern: The case of food safety." *Journal of Public Policy & Marketing*, **11**(2), 12–25.

Fleischer A, Felsenstein D (2002). "Cost-benefit analysis using economic surpluses: A case study of a televised event." *Journal of Cultural Economics*, **26**, 139–156.

Flynn TN (2010). "Valuing citizen and patient preferences in health: Recent developments in three types of best–worst scaling." *Expert Review of PharmacoEconomics & Outcomes Research*, **10**(3), 259–267.

Flynn TN, Louviere JJ, Marley AAJ, Coast J, Peters TJ (2008). "Rescaling quality of life values from discrete choice experiments for use as QALYs: A cautionary tale." *Population Health Metrics*, **6**(6). URL `http://www.pophealthmetrics.com/content/6/1/6`.

Flynn TN, Louviere JJ, Peters TJ, Coast J (2007). "Best–worst scaling: What it can do for health care research and how to do it." *Journal of Health Economics*, **26**, 171–189.

Flynn TN, Louviere JJ, Peters TJ, Coast J (2010). "Using discrete choice experiments to understand preferences for quality of life. Variance-scale heterogeneity matters." *Social Science & Medicine*, **70**, 1957–1965.

Fox J (2005). "The RCommander: A basic-statistics graphical user interface to R." *Journal of Statistical Software*, **14**(9), 1–42. URL `http://www.jstatsoft.org/v14/i09`.

Fox J, Nie Z, Byrnes J (2012). ***sem**: Structural Equation Models, R Package Version 3.0-0.* URL `http://CRAN.R-project.org/package=sem`.

Freeman III AM (1993). *The Measurement of Environmental and Resource Values: Theory and Methods.* Resources for the Future, Washington DC, USA.

Fu TT, Liu JT, Hammitt JK (1999). "Consumer willingness to pay for low-pesticide fresh produce in Taiwan." *Journal of Agricultural Economics*, **50**(2), 220–233.

Gerard K, Shanahan M, Louviere J (2003). "Using stated preference discrete choice modelling to inform health care decision-making: A pilot study of breast screening participation." *Applied Economics*, **35**(9), 1073–1085.

Gerking S, de Haan M, Schulze W (1988). "The marginal value of job safety: A contingent valuation study." *Journal of Risk and Uncertainty*, **1**(2), 185–199.

Goodman S (2009). "An international comparison of retail consumer wine choice." *International Journal of Wine Business Research*, **21**(1), 41–49.

Goto R, Nishimura S, Ida T (2007). "Discrete choice experiment of smoking cessation behaviour in Japan." *Tobacco Control*, **16**(5), 336–343.

Green PE, Krieger AM, Wind Y (2004). "Thirty years of conjoint analysis: Reflections and prospects." In Y Wind, PE Green (eds.), *Marketing Research and Modeling: Progress and Prospects*, pp. 117–139. Kluwer Academic Publishers, MA, USA.

Greene WH (2002). *Econometric Analysis.* 5th edition. Prentice Hall, NJ, USA.

Greene WH, Hensher DA, Rose J (2006). "Accounting for heterogeneity in the variance of unobserved effects in mixed logit models." *Transportation Research Part B: Methodological*, **40**(1), 75–92.

Grimsrud KM, McCluskey JJ, Loureiro ML, Wahl TI (2004). "Consumer attitudes to genetically modified food in Norway." *Journal of Agricultural Economics*, **55**(1), 75–90.

Grömping U (2014a). "CRAN's task view: Design of experiments (DoE) & analysis of experimental data." URL `http://CRAN.R-project.org/web/views/ExperimentalDesign.html`.

Grömping U (2014b). ***DoE.base****: Full Factorials, Orthogonal Arrays and Base Utilities for DoE Packages, R Package Version 0.26.* URL `http://CRAN.R-project.org/package=DoE.base`.

Gustafsson-Wright E, Asfaw A, van der Gaag J (2009). "Willingness to pay for health insurance: An analysis of the potential market for new low-cost health insurance products in Namibia." *Social Science & Medicine*, **69**, 1351–1359.

Haefeli M, Elfering A, McIntosh E, Gray A, Sukthankar A, Boos N (2008). "A cost-benefit analysis using contingent valuation techniques: A feasibility study in spinal surgery." *Value in Health*, **11**, 575–588.

Hall J, Kenny P, King M, Louviere J, Viney R, Yeoh A (2002). "Using stated preference discrete choice modelling to evaluate the introduction of varicella vaccination." *Health Economics*, **11**(5), 457–465.

Hanemann WM (1984). "Welfare evaluations in contingent valuation experiments with discrete responses." *American Journal of Agricultural Economics*, **66**(3), 332–341.

Hanemann WM (1985). "Some issues in continuous- and discrete-response contingent valuation studies." *Northeastern Journal of Agricultural Economics*, **14**, 5–13.

Hanemann WM, Kanninen B (1999). "The statistical analysis of discrete-response CV data." In IJ Bateman, KG Willis (eds.), *Valuing Environmental Preferences: Theory and Practice of the Contingent Valuation Methods in the US, EU, and Developing Countries*, pp. 302–441. Oxford University Press, New York.

Hanemann WM, Loomis JB, Kanninen BJ (1991). "Statistical efficiency of double-bounded dichotomous choice contingent valuation." *American Journal of Agricultural Economics*, **73**(4), 1255–1263.

Hanley N, Wright RE, Koop G (2002). "Modelling recreation demand using choice experiments: Climbing in Scotland." *Environmental and Resource Economics*, **22**, 449 – 466.

Hansen TB (1997). "The willingness-to-pay for the Royal Theatre in Copenhagen as a public good." *Journal of Cultural Economics*, **21**(1), 1–28.

Harless DW, Allen FR (1999). "Using the contingent valuation method to measure patron benefits of reference desk service in an academic library." *College and Research Libraries*, **60**(1), 56–69.

Hedayat AS, Sloane NJA, Stufken J (1999). *Orthogonal Arrays: Theory and Applications*. Springer-Verlag, New York.

Hein KA, Jaeger SR, Carr BT, Delahunty CM (2008). "Comparison of five common acceptance and preference methods." *Food Quality and Preference*, **19**(7), 651–661.

Hensher DA, Rose JM, Greene WH (2005). *Applied Choice Analysis: A Primer*. Cambridge University Press, Cambridge, UK.

Hensher DA, Sullivan C (2003). "Willingness to pay for road curviness and road type." *Transportation Research Part D: Transport and Environment*, **8**(2), 139–155.

Hole AR (2007). "A comparison of approaches to estimating confidence intervals for willingness to pay measures." *Health Economics*, **16**, 827–840.

Holmes TP, Adamowicz WL (2003). "Attribute-based methods." In PA Champ, KJ Boyle, TC Brown (eds.), *A Primer on Nonmarket Valuation*, pp. 171–219. Kluwer Academic Publishers, Dordrecht, NL.

Holmes TP, Bergstrom JC, Huszar E, Kask SB, Orr III F (2004). "Contingent valuation, net marginal benefits, and the scale of riparian ecosystem restoration." *Ecological Economics*, **49**, 19–30.

Hu W, Woods T, Bastin S, Cox L, You W (2011). "Assessing consumer willingness to pay for value-added blueberry products using a payment card survey." *Journal of Agricultural and Applied Economics*, **43**(2), 243–258.

Hultkrantz L, Lindberg G, Andersson C (2006). "The value of improved road safety." *Journal of Risk and Uncertainty*, **32**(2), 151–170.

Hunt JD, Stefan KJ (2007). "Tour-based microsimulation of urban commercial movements." *Transportation Research Part B*, **41**, 981–1013.

Ida T, Goto R (2009). "Interdependency among addictive behaviours and time/risk preferences: Discrete choice model analysis of smoking, drinking, and gambling." *Journal of Economic Psychology*, **30**, 608–621.

Imai K, King G, Lau O (2008). "Toward a common framework for statistical analysis and development." *Journal of Computational and Graphical Statistics*, **17**(4), 892–913.

Imai K, King G, Lau O (2012). ***Zelig**: Everyone's Statistical Software, R Package Version 3.5.5.* URL `http://CRAN.R-project.org/package=Zelig`.

Imai K, van Dyk DA (2011). ***MNP**: R Package for Fitting the Multinomial Probit Model, R Package Version 2.6-3.* URL `http://CRAN.R-project.org/package=MNP`.

Jaeger SR, Cardello AV (2009). "Direct and indirect hedonic scaling methods: A comparison of the labeled affective magnitude (LAM) scale and best–worst scaling." *Food Quality and Preference*, **20**(3), 249–258.

Jaeger SR, Danaher PJ, Brodie RJ (2009). "Wine purchase decisions and consumption behaviours: Insights from a probability sample drawn in Auckland, New Zealand." *Food Quality and Preference*, **20**(4), 312–319.

Jaeger SR, Jørgensen AS, Aaslyng MD, Bredie WL (2008). "Best–worst scaling: An introduction and initial comparison with monadic rating for preference elicitation with food products." *Food Quality and Preference*, **19**(6), 579–588.

Jaeger SR, Rose JM (2008). "Stated choice experimentation, contextual influences and food choice: A case study." *Food Quality and Preference*, **19**(6), 539–564.

Johnson FR, Kanninen B, Bingham M, Özdemir S (2007). "Experimental design for stated choice studies." In BJ Kanninen (ed.), *Valuing Environmental Amenities Using Stated Choice Studies: A Common Sense Approach to Theory and Practice*, pp. 159–202. Springer, Dordrecht, NL.

Johnson FR, Lancsar E, Marshall D, Kilambi V, Mühlbacher A, Regier DA, Bresnahan BW, Kanninen B, Bridges JFP (2013). "Constructing experimental designs for discrete-choice experiments: Report of the ISPOR conjoint analysis experimental design good research practices task force." *Value in Health*, **16**, 3–13.

Johnston RJ, Duke JM (2007). "Willingness to pay for agricultural land preservation and policy process attributes: Does the method matter?" *American Journal of Agricultural Economics*, **89**(4), 1098–1115.

Kallas Z, Gómez-Limón JA, Arriaza M (2007). "Are citizens willing to pay for agricultural multifunctionality?" *Agricultural Economics*, **36**(3), 405–419.

Kanninen BJ (1993a). "Design of sequential experiments for contingent valuation studies." *Journal of Environmental Economics and Management*, **25**(1), S1–S11.

Kanninen BJ (1993b). "Optimal experimental design for double-bounded dichotomous choice contingent valuation." *Land Economics*, **69**(2), 138–146.

Kanninen BJ (ed.) (2007). *Valuing Environmental Amenities Using Stated Choice Studies: A Common Sense Approach to Theory and Practice.* Springer, Dordrecht, NL.

Karousakis K, Birol E (2008). "Investigating household preferences for kerbside recycling services in London: A choice experiment approach." *Journal of Environmental Management*, **88**(4), 1099–1108.

Kim D, Canh DG, Poulos C, Thoa LTK, Cook J, Hoa NT, Nyamete A, Thuy DTD, Deen J, Clemens J, Thiem VD, Anh DD, Whittington D (2008). "Private demand for cholera vaccines in Hue, Vietnam." *Value in Health*, **11**(1), 119–128.

Kimura A, Wada Y, Kamada A, Masuda T, Okamoto M, Goto S, Tsuzuki D, Cai D, Oka T, Dan I (2010). "Interactive effects of carbon footprint information and its accessibility on value and subjective qualities of food products." *Appetite*, **55**(2), 271–278.

Kleiber C, Zeileis A (2008). *Applied Econometrics with R.* Springer, New York.

Kling RW, Revier CF, Sable K (2004). "Estimating the public good value of preserving a local historic landmark: The role of non-substitutability and citizen information." *Urban Studies*, **41**(10), 2025–2041.

Kolstad JR (2011). "How to make rural jobs more attractive to health workers. Findings from a discrete choice experiment in Tanzania." *Health Economics*, **20**(2), 196–211.

Krinsky I, Robb AL (1986). "On approximating the statistical properties of elasticities." *The Review of Economics and Statistics*, **68**, 715–719.

Krinsky I, Robb AL (1990). "On approximating the statistical properties of elasticities: A correction." *The Review of Economics and Statistics*, **72**, 189–190.

Kriström B (1990). "A non-parametric approach to the estimation of welfare measures in discrete response valuation studies." *Land Economics*, **66**(2), 135–139.

Kruger C, Boxall PC, Luckert MK (2013). "Preferences of community public advisory group members for characteristics of Canadian forest tenures in pursuit of sustainable forest management objectives." *Forest Policy and Economics*, **26**, 121–130.

Krupnick A, Alberini A, Cropper M, Simon N, O'Brien B, Goeree R, Heintzelman M (2002). "Age, health, and the willingness to pay for mortality risk reductions: A contingent valuation survey of Ontario residents." *Journal of Risk and Uncertainty*, **24**(2), 161–186.

Kuhfeld WF (2006). "Orthogonal arrays." *SAS Technical Paper*, **TS-723**. URL `http://support.sas.com/techsup/technote/ts723.html`.

Lagerkvist CJ, Carlsson F, Viske D (2006). "Swedish consumer preferences for animal welfare and biotech: A choice experiment." *AgBioForum*, **9**(1), 51–58. URL `http://www.agbioforum.org`.

Lancsar E, Louviere J (2008). "Conducting discrete choice experiments to inform healthcare decision making: A user's guide." *PharmacoEconomics*, **26**(8), 661–677.

Lancsar E, Louviere J, Donaldson C, Currie G, Burgess L (2013). "Best worst discrete choice experiments in health: Methods and an application." *Social Science & Medicine*, **76**, 74–82.

Lang HC (2005). "Economic grand rounds: Patients' and caregivers' willingness to pay for a cure for schizophrenia in Taiwan." *Psychiatric Services*, **56**(2), 149–151.

Lang HC (2010). "Willingness to pay for lung cancer treatment." *Value in Health*, **13**(6), 743–749.

Laver K, Ratcliffe J, George S, Burgess L, Crotty M (2011). "Is the Nintendo Wii Fit really acceptable to older people?: A discrete choice experiment." *BMC Geriatrics*, **11**(64). URL `http://www.biomedcentral.com/1471-2318/11/64`.

Lee JA, Soutar G, Louviere J (2008). "The Best–worst scaling approach: An alternative to Schwartz's values survey." *Journal of Personality Assessment*, **90**(4), 335–347.

Lee JA, Soutar GN, Louviere J (2007). "Measuring values using best–worst scaling: The LOV example." *Psychology & Marketing*, **24**(12), 1043–1058.

Leitham S, McQuaid RW, Nelson JD (2000). "The influence of transport on industrial location choice: A stated preference experiment." *Transportation Research Part A: Policy and Practice*, **34**(7), 515–535.

Li Q, Curtis KR, McCluskey JJ, Wahl TI (2002). "Consumer attitudes toward genetically modified foods in Beijing, China." *AgBioForum*, **5**(4), 145–152. URL `http://www.agbioforum.org`

Lieu TA, Ortega-Sanchez I, Ray GT, Rusinak D, Yih WK, Choo PW, Shui I, Kleinman K, Harpaz R, Prosser LA (2008). "Community and patient values for preventing herpes zoster." *PharmacoEconomics*, **26**(3), 235–249.

Lockwood M, Tracey P, Klomp N (1996). "Analysing conflict between cultural heritage and nature conservation in the Australian Alps: A CVM approach." *Journal of Environmental Planning and Management*, **39**(3), 357–370.

Longworth L, Ratcliffe J, Boulton M (2001). "Investigating women's preferences for intrapartum care: Home versus hospital births." *Health & Social Care in the Community*, **9**(6), 404–413.

Loomis J, Kent P, Strange L, Fausch K, Covich A (2000). "Measuring the total economic value of restoring ecosystem services in an impaired river basin: Results from a contingent valuation survey." *Ecological Economics*, **33**(1), 103–117.

Loureiro ML (2003). "Rethinking new wines: Implications of local and environmentally friendly labels." *Food Policy*, **28**, 547–560.

Loureiro ML, Arcos FD (2012). "Applying best–worst scaling in a stated preference analysis of forest management programs." *Journal of Forest Economics*, **18**(4), 381–394.

Loureiro ML, Gracia A, Nayga RM (2006). "Do consumers value nutritional labels?" *European Review of Agricultural Economics*, **33**(2), 249–268.

Loureiro ML, McCluskey JJ, Mittelhammer RC (2002). "Will consumers pay a premium for eco-labeled apples?" *Journal of Consumer Affairs*, **36**(2), 203–219.

Louviere J, Lings I, Islam T, Gudergan S, Flynn T (2013). "An introduction to the application of (case 1) best–worst scaling in marketing research." *International Journal of Research in Marketing*, **30**, 292–303.

Louviere JJ, Fiebig DG (2010). "Benefit assessment for cost-benefit analysis studies in health care using discrete choice experiments: Estimating welfare in a health care setting." In E McIntosh, PM Clarke, EJ Frew, JJ Louviere (eds.), *Applied Methods of Cost-Benefit Analysis in Health Care*, pp. 211–229. Oxford University Press, New York.

Louviere JJ, Flynn TN (2010). "Using best–worst scaling choice experiments to measure public perceptions and preferences for healthcare reform in Australia." *The Patient: Patient-Centered Outcomes Research*, **3**(4), 275–283.

Louviere JJ, Flynn TN, Carson RT (2010). "Discrete choice experiments are not conjoint analysis." *Journal of Choice Modelling*, **3**(3), 57–72.

Louviere JJ, Hensher DA, Swait JD (2000). *Stated Choice Methods: Analysis and Application.* Cambridge University Press, Cambridge, UK.

Louviere JJ, Islam T (2008). "A comparison of importance weights and willingness-to-pay measures derived from choice-based conjoint, constant sum scales and best–worst scaling." *Journal of Business Research*, **61**, 903–911.

Louviere JJ, Street D, Burgess L, Wasi N, Islam T, Marley AAJ (2008). "Modeling the choices of individual decision-makers by combining efficient choice experiment designs with extra preference information." *Journal of Choice Modelling*, **1**(1), 128–163.

Louviere JJ, Woodworth G (1983). "Design and analysis of simulated consumer choice or allocation experiments: An approach based on aggregate data." *Journal of Marketing Research*, **20**, 350–367.

Ludwig J, Cook PJ (2001). "The benefits of reducing gun violence: Evidence from contingent-valuation survey data." *Journal of Risk and Uncertainty*, **22**(3), 207–226.

Lusk JL (2003). "Effects of cheap talk on consumer willingness-to-pay for golden rice." *American Journal of Agricultural Economics*, **85**(4), 840–856.

Lusk JL, Briggeman BC (2009). "Food values." *American Journal of Agricultural Economics*, **91**(1), 184–196.

Lusk JL, Fox JA (2002). "Consumer demand for mandatory labeling of beef from cattle administered growth hormones or fed genetically modified corn." *Journal of Agricultural and Applied Economics*, **34**, 27–38.

Lusk JL, Marette S (2010). "Welfare effects of food labels and bans with alternative willingness to pay measures." *Applied Economic Perspectives and Policy*, **32**(2), 319–337.

Lusk JL, Nilsson T, Foster K (2007). "Public preferences and private choices: Effect of altruism and free riding on demand for environmentally certified pork." *Environmental and Resource Economics*, **36**, 499–521.

Lusk JL, Parker N (2009). "Consumer preferences for amount and type of fat in ground beef." *Journal of Agricultural and Applied Economics*, **41**(1), 75–90.

Lusk JL, Schroeder TC (2004). "Are choice experiments incentive compatible? A test with quality differentiated beef steaks." *American Journal of Agricultural Economics*, **86**(2), 467–482.

Macmillan DC, Philip L, Hanley N, Alvarez-Farizo B (2002). "Valuing the non-market benefits of wild goose conservation: A comparison of interview and group-based approaches." *Ecological Economics*, **43**, 49–59.

Madureira L, Rambonilaza T, Karpinski I (2007). "Review of methods and evidence for economic valuation of agricultural non-commodity outputs and suggestions to facilitate its application to broader decisional contexts." *Agriculture, Ecosystems & Environment*, **120**, 5–20.

Maechler M, Rousseeuw P, Struyf A, Hubert M, Hornik K (2012). ***cluster**: Cluster Analysis Basics and Extensions, R Package Version 1.14-2.* URL `http://CRAN.R-project.org/package=cluster`.

Mallawaarachchi T, Quiggin J (2001). "Modelling socially optimal land allocations for sugar cane growing in North Queensland: A linked mathematical programming and choice modelling study." *The Australian Journal of Agricultural and Resource Economics*, **45**(3), 383–409.

Mangione TW (1995). *Mail Surveys: Improving the Quality.* Sage Publications, CA, USA.

Marley AAJ (2010). "The best–worst method for the study of preferences." In PA Frensch, R Schwarzer (eds.), *Cognition and Neuropsychology: International Perspectives on Psychological Science*, pp. 147–157. Psychology Press, Hove, UK.

Marley AAJ, Flynn TN, Louviere JJ (2008). "Probabilistic models of set-dependent and attribute-level best–worst choice." *Journal of Mathematical Psychology*, **52**, 281–296.

Marley AAJ, Louviere JJ (2005). "Some probabilistic models of best, worst, and best–worst choices." *Journal of Mathematical Psychology*, **49**, 464–480.

Marley AAJ, Pihlens (2012). "Models of best–worst choice and ranking among multiattribute options (profiles)." *Journal of Mathematical Psychology*, **56**, 24–34.

Marti J (2012). "A best–worst scaling survey of adolescents' level of concern for health and non-health consequences of smoking." *Social Science & Medicine*, **75**(1), 87–97.

Martin F (1994). "Determining the size of museum subsidies." *Journal of Cultural Economics*, **18**(4), 255–270.

Masiero L, Hensher DA (2010). "Analyzing loss aversion and diminishing sensitivity in a freight transport stated choice experiment." *Transportation Research Part A: Policy and Practice*, **44**(5), 349–358.

Mazanov J, Huybers T, Connor J (2012). "Prioritising health in anti-doping: What Australians think." *Journal of Science and Medicine in Sport*, **15**(5), 381–385.

McDaniels TL (1992). "Reference points, loss aversion, and contingent values for auto safety." *Journal of Risk and Uncertainty*, **5**(2), 187–200.

McIntosh E, Ryan M (2002). "Using discrete choice experiments to derive welfare estimates for the provision of elective surgery: Implications of discontinuous preferences." *Journal of Economic Psychology*, **23**(3), 367–382.

McNair BJ, Bennett J, Hensher DA (2011). "A comparison of responses to single and repeated discrete choice questions." *Resource and Energy Economics*, **33**(3), 554–571.

Menictas C, Wang PZ, Louviere JJ (2012). "Assessing the validity of brand equity constructs." *Australasian Marketing Journal*, **20**(1), 3–8.

Mielby LH, Edelenbos M, Thybo AK (2012). "Comparison of rating, best–worst scaling, and adolescents' real choices of snacks." *Food Quality and Preference*, **25**(2), 140–147.

Mitchell RC, Carson RT (1989). *Using Surveys to Value Public Goods: The Contingent Valuation Method.* Resources for the Future, Washington, D.C.

Molin E, Oppewal H, Timmermans H (1996). "Predicting consumer response to new housing: A stated choice experiment." *Netherlands Journal of Housing and the Built Environment*, **11**, 297–311.

Molin EJE, Timmermans HJP (2010). "Context dependent stated choice experiments: The case of train egress mode choice." *Journal of Choice Modelling*, **3**(3), 39–56.

Morris MD (2011). *Design of Experiments: An Introduction Based on Liner Models.* Chapman & Hall/CRC, Boca Raton, FL.

Morrison M, Bennett J, Blamey R, Louviere J (2002). "Choice modeling and tests of benefit transfer." *American Journal of Agricultural Economics*, **84**(1), 161–170.

Mueller S, Francis IL, Lockshin L (2009). "Comparison of best–worst and hedonic scaling for the measurement of consumer wine preferences." *Australian Journal of Grape and Wine Research*, **15**, 205–215.

Mueller S, Lockshin L, Louviere JJ (2010). "What you see may not be what you get: Asking consumers what matters may not reflect what they choose." *Marketing Letters*, **21**, 335–350.

Mueller S, Rungie C (2009). "Is there more information in best worst choice data?: Using the attitude heterogeneity structure to identify consumer segments." *International Journal of Wine Business Research*, **21**(1), 24–40.

Mueller Loose S, Lockshin L (2013). "Testing the robustness of best worst scaling for cross-national segmentation with different numbers of choice sets." *Food Quality and Preference*, **27**(2), 230–242.

Nakatani T (2014). ***DCchoice**: A Package for Analyzing Dichotomous Choice Contingent Valuation Data.* R Package Version 0.0.6-5, URL `https://r-forge.r-project.org/projects/dcchoice/`.

Nam KM, Selin NE, Reilly JM, Paltsev S (2010). "Measuring welfare loss caused by air pollution in Europe: A CGE analysis." *Energy Policy*, **38**, 5059–5071.

Nordström J (2012). "Willingness to pay for wholesome canteen takeaway." *Appetite*, **58**, 168–179.

O'Brien BJ, Novosel S, Torrance G, Streiner D (1995). "Assessing the economic value of a new antidepressant: A willingness-to-pay approach." *PharmacoEconomics*, **8**(1), 34–45.

OECD (2001). *Multifunctionality: Towards an Analytical Framework.* OECD, Paris, France.

Onozaka Y, McFadden DT (2011). "Does local labeling complement or compete with other sustainable labels? A conjoint analysis of direct and joint values for fresh produce claim." *American Journal of Agricultural Economics*, **93**(3), 693–706.

Orme B (2005). "Accuracy of HB estimation in MaxDiff experiments." *Sawtooth Software Research Paper Series*, pp. 1–7.

Orme BK (2010). *Getting Started with Conjoint Analysis: Strategies for Product Design and Pricing Research.* Second edition. Research Publishers LLC, WI, USA.

Pek CK, Jamal O (2011). "A choice experiment analysis for solid waste disposal option: A case study in Malaysia." *Journal of Environmental Management*, **92**(11), 2993–3001.

Persson U, Norinder A, Hjalte K, Gralén K (2001). "The value of a statistical life in transport: Findings from a new contingent valuation study in Sweden." *Journal of Risk and Uncertainty*, **23**(2), 121–134.

Poe GL, Giraud KL, Loomis JB (2005). "Computational methods for measuring the difference of empirical distributions." *American Journal of Agricultural Economics*, **87**(2), 353–365.

Pollicino M, Maddison D (2001). "Valuing the benefits of cleaning Lincoln Cathedral." *Journal of Cultural Economics*, **25**(2), 131–148.

Potoglou D, Burge P, Flynn T, Netten A, Malley J, Forder J, Brazier JE (2011). "Best–worst scaling vs. discrete choice experiments: An empirical comparison using social care data." *Social Science & Medicine*, **72**, 1717–1727.

Puckett SM, Hensher DA (2009). "Revealing the extent of process heterogeneity in choice analysis: An empirical assessment." *Transportation Research Part A: Policy and Practice*, **43**(2), 117–126.

Raghavarao D, Wiley JB, Chitturi P (2011). *Choice-Based Conjoint Analysis: Models and Designs.* Chapman & Hall/CRC, Boca Raton, FL.

Randall A (2002). "Valuing the outputs of multifunctional agriculture." *European Review of Agricultural Economics*, **29**, 289–307.

R Core Team (2014). *R: A Language and Environment for Statistical Computing.* R Foundation for Statistical Computing, Vienna, Austria. URL `http://www.R-project.org/`.

R Core Team (2014). *R Data Import/Export.* R Foundation for Statistical Computing, Vienna, Austria. URL `http://www.R-project.org/`.

R Core Team (2014). *R Language Definitions.* R Foundation for Statistical Computing, Vienna, Austria. URL `http://www.R-project.org/`.

Remaud H, Lockshin L (2009). "Building brand salience for commodity-based wine regions." *International Journal of Wine Business Research*, **21**(1), 79–92.

Renting H, Rossing WAH, Groot JCJ, Van der Ploeg JD, Laurent C, Perraud D, Stobbelaar DJ, Van Ittersum MK (2009). "Exploring multifunctional agriculture. A review of conceptual approaches and prospects for an integrative transitional framework." *Journal of Environmental Management*, **90**, S112–S123.

Revelle W (2012). ***psych**: Procedures for Psychological, Psychometric, and Personality Research.* Northwestern University, Il, USA. R package version 1.2.1, URL `http://personality-project.org/r/psych.manual.pdf`.

Rivers N, Jaccard M (2005). "Combining top-down and bottom-up approaches to energy-economy modeling using discrete choice methods." *The Energy Journal*, **26**(1), 83–106.

Roe B, Teisl MF, Levy A, Russell M (2001). "US consumers' willingness to pay for green electricity." *Energy Policy*, **29**(11), 917–925.

Rolfe J, Bennett J (eds.) (2006). *Choice Modelling and the Transfer of Environmental Values.* Edward Elgar, Cheltenham, UK.

Roosen J, Bieberstein A, Marette S, Blanchemanche S, Vandermoere F (2011). "The effect of information choice and discussion on consumers' willingness-to-pay for nanotechnologies in food." *Journal of Agricultural and Resource Economics*, **36**(2), 365–374.

Rossi P (2012). ***bayesm**: Bayesian Inference for Marketing/Micro-Econometrics, R Package Version 2.2-5.* URL `http://CRAN.R-project.org/package=bayesm`.

Rossi PE, Allenby GM, McCulloch R (2005). *Bayesian Statistics and Marketing.* John Wiley & Sons, West Sussex, UK.

Rousseau S, Vranken L (2013). "Green market expansion by reducing information asymmetries: Evidence for labeled organic food products." *Food Policy*, **40**, 31–43.

Rudd MA (2011). "Scientists' opinions on the global status and management of biological diversity." *Conservation Biology*, **25**(6), 1165–1175.

Ryan M (1999). "Using conjoint analysis to take account of patient preferences and go beyond health outcomes: An application to in vitro fertilisation." *Social Science & Medicine*, **48**(4), 535–546.

Ryan M, Gerard K, Amaya-Amaya M (eds.) (2008a). *Using Discrete Choice Experiments to Value Health and Health Care.* Springer, Dordrecht, NL.

Ryan M, Netten A, Skåtun D, Smith P (2006). "Using discrete choice experiments to estimate a preference-based measure of outcome: An application to social care for older people." *Journal of Health Economics*, **25**, 927–944.

Ryan M, Watson V, Gerard K (2008b). "Practical issues in conducting a discrete choice experiment." In M Ryan, K Gerard, M Amaya-Amaya (eds.), *Using Discrete Choice Experiments to Value Health and Health Care*, pp. 73–97. Springer, Dordrecht, NL.

Sailer MO (2013). ***crossdes**: Design and Randomization in Crossover Studies, R Package Version 1.1-1.* URL `http://CRAN.R-project.org/package=crossdes`.

Santagata W, Signorello G (2000). "Contingent valuation of a cultural public good and policy design: The case of 'Napoli Musei Aperti'." *Journal of Cultural Economics*, **24**(3), 181–204.

Sanz JÁ, Herrero LC, Bedate AM (2003). "Contingent valuation and semi-parametric methods: A case study of the National Museum of Sculpture in Valladolid, Spain." *Journal of Cultural Economics*, **27**, 241–257.

Scarpa R, Rose JM (2008). "Design efficiency for non-market valuation with choice modelling: How to measure it, what to report and why." *The Australian Journal of Agricultural and Resource Economics*, **52**, 253–282.

Scarpa R, Willis KG, Acutt M (2007). "Valuing externalities from water supply: Status quo, choice complexity and individual random effects in panel kernel logit analysis of choice experiments." *Journal of Environmental Planning and Management*, **50**, 449–466.

Shapansky B, Adamowicz WL, Boxall PC (2008). "Assessing information provision and respondent involvement effects on preferences." *Ecological Economics*, **65**(3), 626–635.

Sirieix L, Remaud H, Lockshin L, Thach L, Lease T (2011). "Determinants of restaurant's owners/managers selection of wines to be offered on the wine list." *Journal of Retailing and Consumer Services*, **18**(6), 500–508.

Sloane NJA (2014). "A library of orthogonal arrays." URL `http:/nellsloane.com/oadir/`.

Smith VK (2006). "Fifty years of contingent valuation." In A Alberini, JR Kahn (eds.), *Handbook on Contingent Valuation*, pp. 7–65. Edward Elgar, Northampton, MA.

Street DJ, Burgess L (2007). *The Construction of Optimal Stated Choice Experiments: Theory and Methods.* John Wiley & Sons, NJ, USA.

Susaeta A, Alavalapati J, Lal P, Matta JR, Mercer E (2010). "Assessing public preferences for forest biomass based energy in the Southern United States." *Environmental Management*, **45**, 697–710.

Takama T, Preston J (2008). "Forecasting the effects of road user charge by stochastic agent-based modelling." *Transportation Research Part A*, **42**, 738–749.

Tappenden P, Brazier J, Ratcliffe J, Chilcott J (2007). "A stated preference binary choice experiment to explore NICE decision making." *PharmacoEconomics*, **25**(8), 685–693.

Tavares S, Cardoso M, Dias JG (2010). "The heterogeneous best–worst choice method in market research." *International Journal of Market Research*, **52**, 533–546.

Telser H, Zweifel P (2002). "Measuring willingness-to-pay for risk reduction: An application of conjoint analysis." *Health Economics*, **11**(2), 129–139.

Therneau T (2014). ***survival**: Survival Analysis, Including Penalised Likelihood, R Package Version 2.37-7.* URL `http://CRAN.R-project.org/package=survival`.

Thomson DMH, Crocker C, Marketo CG (2010). "Linking sensory characteristics to emotions: An example using dark chocolate." *Food Quality and Preference*, **21**(8), 1117–1125.

Tonsor GT, Shupp R (2009). "Valuations of 'Sustainably Produced' labels on beef, tomato, and apple products." *Agricultural and Resource Economics Review*, **38**(3), 371–383.

Tonsor GT, Wolf CA (2011). "On mandatory labeling of animal welfare attributes." *Food Policy*, **36**, 430–437.

Toomet O (2012). ***intReg**: Interval Regression, R Package Version 0.1-2.* URL `http://CRAN.R-project.org/package=intReg`.

Torbica A, Fattore G (2010). "Understanding the impact of economic evidence on clinical decision making: A discrete choice experiment in cardiology." *Social Science & Medicine*, **70**(10), 1536–1543.

Train KE (1986). *Qualitative Choice Analysis: Theory, Econometrics, and an Application to Automobile Demand.* The MIT Press, MA, USA.

Train KE (2009). *Discrete Choice Methods with Simulation.* Second edition. Cambridge University Press, NY, USA.

Turnbull BW (1976). "The empirical distribution function with arbitrarily grouped, censored and truncated data." *Journal of the Royal Statistical Society. Series B*, **38**(3), 290–295.

Tutz G (2011). *Regression for Categorical Data.* Cambridge University Press, New York, USA.

Unterschultz J, Quagrainie KK, Veeman M, Kim RB (1998). "South Korean hotel meat buyers' perceptions of Australian, Canadian and U.S. beef." *Canadian Journal of Agricultural Economics/Revue canadienne d'agroeconomie*, **46**(1), 53–68.

van den Berg B, Bleichrodt H, Eeckhoudt L (2005). "The economic value of informal care: A study of informal caregivers' and patients' willingness to pay and willingness to accept for informal care." *Health Economics*, **14**(4), 363–376.

van Hulst LTC, Kievit W, van Bommel R, van Riel PLCM, Fraenkel L (2011). "Rheumatoid arthritis patients and rheumatologists approach the decision to escalate care differently: Results of a maximum difference scaling experiment." *Arthritis Care & Research*, **63**(10), 1407–1414.

Venables WN, Ripley BD (2002). *Modern Applied Statistics with S*. Fourth edition. Springer, New York. ISBN 0-387-95457-0, URL `http://www.stats.ox.ac.uk/pub/MASS4`.

Venables WN, Smith DM, the R Core Team (2014). *An Introduction to R*. R Foundation for Statistical Computing, Vienna, Austria. URL `http://www.R-project.org/`.

Verbeek M (2004). *A Guide to Modern Econometrics*. John Wiley & Sons, NY, USA.

Wang T, Wong B, Huang A, Khatri P, Ng C, Forgie M, Lanphear JH, O'Neill PJ (2011). "Factors affecting residency rank-listing: A Maxdiff survey of graduating Canadian medical students." *BMC Medical Education*, **11**(61). URL `http://www.biomedcentral.com/1472-6920/11/61`.

Weaver M, Ndamobissi R, Kornfield R, Blewane C, Sathe A, Chapko M, Bendje N, Nguembi E, Senwara-Defiobonna J (1996). "Willingness to pay for child survival: Results of a national survey in Central African Republic." *Social Science & Medicine*, **43**(6), 985–998.

Wheeler B (2014). ***AlgDesign**: Algorithmic Experimental Design, R Package Version 1.1-7.2*. URL `http://CRAN.R-project.org/package=AlgDesign`.

Whitehead JC (1995). "Willingness to pay for quality improvements: Comparative statics and interpretation of contingent valuation results." *Land Economics*, **71**(2), 207–215.

Whitehead JC, Finney SS (2003). "Willingness to pay for submerged maritime cultural resources." *Journal of Cultural Economics*, **27**, 231–240.

Whitehead JC, Haab TC, Huang JC (1998). "Part-whole bias in contingent valuation: Will scope effects be detected with inexpensive survey methods?" *Southern Economic Journal*, **65**(1), 160–168.

Whittington D, Matsui-Santana O, Freiberger JJ, Houtven GV, Pattanayak S (2002). "Private demand for a HIV/AIDS vaccine: Evidence from Guadalajara, Mexico." *Vaccine*, **20**, 2585–2591.

Witt J, Scott A, Osborne RH (2009). "Designing choice experiments with many attributes. An application to setting priorities for orthopaedic waiting lists." *Health Economics*, **18**(6), 681–696.

Wooldridge JM (2002). *Econometric Analysis of Cross Section and Panel Data*. The MIT Press, MA, USA.

Yee TW (2012). ***VGAM**: Vector Generalized Linear and Additive Models, R Package Version 0.9-0*. URL `http://CRAN.R-project.org/package=VGAM`.

Yee TW, Wild CJ (1996). "Vector generalized additive models." *The Journal of Royal Statistical Society, Series B*, **58**(3), 481–493.

Ying XH, Hu TW, Ren J, Chen W, Xu K, Huang JH (2007). "Demand for private health insurance in Chinese urban areas." *Health Economics*, **16**(10), 1041–1050.

Zethraeus N (1998). "Willingness to pay for hormone replacement therapy." *Health Economics*, **7**, 31–38.

Index